brilliant
eBay®

Dom Brookman

Harlow, England • London • New York • Boston • San Francisco • Toronto
Sydney • Tokyo • Singapore • Hong Kong • Seoul • Taipei • New Delhi
Cape Town • Madrid • Mexico City • Amsterdam • Munich • Paris • Milan

Pearson Education Limited
Edinburgh Gate
Harlow
Essex CM20 2JE
England

and Associated Companies throughout the world

Visit us on the World Wide Web at:
www.pearsoned.co.uk

First published 2006

© Pearson Education Limited 2006

ISBN-13: 978-0-13-223965-3
ISBN-10: 0-13-223965-5

British Library Cataloguing-in-Publication Data
A catalogue record for this book is available from the British Library

Library of Congress Cataloging-in-Publication Data
A catalog record for this book is available from the Library of Congress

10 9 8 7 6 5 4 3 2 1
10 09 08 07 06

Prepared for Pearson Education Ltd by Syllaba Ltd (http://www.syllaba.co.uk).
Typeset in 8.5pt Helvetica by 30
Printed and bound in Great Britain by Ashford Colour Press Ltd, Gosport.

The publisher's policy is to use paper manufactured from sustainable forests.

Brilliant guides

What you need to know and how to do it

When you come up against a problem that you're unsure how to solve, or want to accomplish something that you aren't sure how to do, where do you look? Manuals and traditional training guides are usually too big and unwieldy and are intended to be used as end-to-end training resources, making it hard to get to the info you need right away without having to wade through pages of background information that you just don't need at that moment – and helplines are rarely that helpful!

Brilliant guides have been developed to allow you to find the info you need easily and without fuss and guide you through the task using a highly visual, step-by-step approach – providing exactly what you need to know when you need it!

Brilliant guides provide the quick easy-to-access information that you need, using a detailed table of contents and troubleshooting guide to help you find exactly what you need to know, and then presenting each task in a visual manner. Numbered steps guide you through each task or problem, using numerous screenshots to illustrate each step. Added features include 'See also' boxes that point you to related tasks and information in the book, while 'Did you know?' sections alert you to relevant expert tips, tricks and advice to further expand your skills and knowledge.

The new *Brilliant Lifestyle* guides bring the *Brilliant* philosophy to your digital lifestyle requirements. New titles on eBay, iPod and Online Poker provide all the quick easy-to-access information you need to take full advantage of the opportunities to transform the way you shop, listen to music and play cards.

In addition to covering all major office PC applications, and related computing subjects, the *Brilliant* series also contains titles that will help you in every aspect of your working life, such as writing the perfect CV, answering the toughest interview questions and moving on in your career.

Brilliant guides are the light at the end of the tunnel when you are faced with any minor or major task.

Author's acknowledgements

Thanks to: Rafa Benitez, David, Rachael and Susan Brookman, Lee Cornish, Tom Evans, Dan Ford, Beth Orton, Kate Phelps, Matt Powell, Robert Powell, Dawn and Rob Neatby, Jennie, Ollie and Steve Sanders, Anita Wyatt.

Dedication

This book is dedicated to Rozalind Neatby.

Publisher's acknowledgements

eBay® is a trademark of eBay, Inc. *Brilliant eBay* is an independent publication and has not been authorised, sponsored, or otherwise approved by eBay, Inc.

The author and publisher would like to thank the following for permission to reproduce the material in this book: Skype Ltd, Auction LotWatch, AuctionBlackList.com, Swift-Find Ltd, Symantec Corp., ChannelAdvisor Corp., Marketworks Inc., Oddcast Corp., HammerTap, Actinic Software Ltd, Dreamteam Design Ltd, 3D3.COM Pty. Ltd, PayPal, Nochex.

Microsoft product screen shots reprinted with permission from Microsoft Corporation.

In some instances we have been unable to trace the owners of copyright material, and we would appreciate any information that would enable us to do so.

About the author

Dom Brookman is a freelance editor and writer. He has edited a number of computer-based titles over the last seven years, including *Internet Made Easy, Internet User, eBuyer* and *Online Seller* and *Word Made Easy*. His first book, entitled *Brilliant Internet*, was published in December 2005. Away from work he enjoys following Liverpool FC, travelling the world and going to music concerts. He is 29 years old and lives in Bournemouth.

Contents

Introduction

HOW YOU'LL LEARN

→ **Find what you need to know – when you need it**

→ **How this book works**

→ **Step-by-step instructions**

→ **Troubleshooting guide**

→ **Spelling**

Welcome to *Brilliant eBay*, a visual quick reference book that shows you how to make the most of one of the world's most popular websites, eBay. Discover the best ways to get the most out of this auction giant, and the tips you need to buy for less, and sell for more.

Find what you need to know – when you need it

You don't have to read this book in any particular order. We've designed the book so that you can jump in, get the information you need, and jump out. To find the information that you need, just look up the task in the table of contents or Troubleshooting guide, and turn to the page listed. Read the task introduction, follow the step-by-step instructions along with the illustration, and you're done.

How this book works

Each task is presented with step-by-step instructions and annotated screen illustrations on the same page. This arrangement lets you focus on a single task without having to turn the pages too often.

Step-by-step instructions

This book provides concise step-by-step instructions that show you how to accomplish a task. Each set of instructions includes illustrations that directly correspond to the easy-to-read steps. Eye-catching text features provide additional helpful information in bite-sized chunks to help you work more efficiently or to teach you more in-depth information. The 'For your information' feature provides tips and techniques to help you work smarter, while the 'See also' cross-references lead you to other parts of the book containing related information about the task. Essential information is highlighted in 'Important' boxes that will ensure you don't miss any vital suggestions and advice.

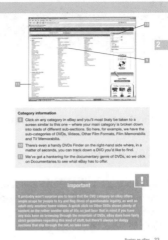

Category information

Click on any category in eBay and you'll most likely be taken to a screen similar to this one – where your main category is broken down into loads of different sub-sections. So here, for example, we have the sub-categories of DVDs, Videos, Other Film Formats, Film Memorabilia and TV Memorabilia.

There's even a handy DVDs Finder on the right-hand side where, in a matter of seconds, you can track down a DVD you'd like to find.

We've got a hankering for the documentary genre of DVDs, so we click on Documentaries to see what eBay has to offer.

Important

It probably won't surprise you to learn that the DVD category on eBay offers ample scope for people to try and flog items of questionable legality, as well as adult-only amateur home videos. A quick click on Other DVDs shows plenty of content on the rather seedier side of life, so just bear that in mind if you have any kids keen on browsing through the mountain of DVDs, eBay does have fairly strict guidelines regarding this kind of stuff, but there'll always be dodgy sections that slip through the net, so take care.

Troubleshooting guide

This book offers quick and easy ways to diagnose and solve common problems that you might encounter using the Troubleshooting guide. The problems are grouped into categories that are presented alphabetically.

Spelling

We have used UK spelling conventions throughout this book. You may therefore notice some inconsistencies between the text and the software on your computer which is likely to have been developed in the USA. We have however adopted US spelling for the words 'disk' and 'program' as these are becoming commonly accepted throughout the world.

1 Introduction to eBay

WHAT YOU'LL DO

→ Register with eBay

→ Find out about eBay's history

→ Learn about eBay's partners

→ Learn about eBay around the world

→ Learn about eBay and Skype

Hello and welcome to *Brilliant eBay*, a new guide to the world's most successful eCommerce site, where millions of people buy and sell online with like-minded users, 24 hours per day, 7 days per week and 365 days per year.

Even if you're a newcomer to the internet, there's little chance that you haven't at the very least heard of eBay and its incredible success. The human-life stories that eBay throws up seemingly every day are manna from heaven for the national press, and barely a month seems to pass by without a new eBay advertising campaign hitting our television screens. To put it bluntly, eBay has been an absolutely staggering success, is one of the most recognisable names both online and in our everyday language, and, it can be argued, has been one of the major players in the huge rise of the internet in our everyday lives over the last 5 years.

eBay's popularity means that there's no point in us denying that, currently in your local bookshop, there are dozens of guides, manuals and weighty tomes dedicated to the site. *Brilliant eBay* is different, however, because we cut through all the hype and jargon, and answer the questions that really matter to you when it comes to using the world's favourite online marketplace. How can you spot a bargain on eBay? What are the best techniques to make your item stand out online? How, to put it bluntly, can you make as much money in as short a time as possible? Is it possible for anyone to become a PowerSeller? How do you set up your own shop?

We'll be aiming to answer these questions and many, many more over the course of the book. We'll also be aiming to dispel a few myths about the site that have been propelled into the national consciousness by the media over the last couple of years or so. Although occasionally you'll read an article that concentrates on some of the amazing statistics that the site generates – how many sales there are on eBay per minute across the world, for example – many reports choose to dwell on some of its negative aspects. Ticket touts, dodgy sellers, even moral and ethical issues all hit the papers during 2005 – witness the furore when it was found out that people were flogging their (free) Live8 tickets on the site to eager punters desperate to see one of the biggest music concerts ever held. Our message is – as in real life, eBay does have its share of unscrupulous users, but with the right mixture of common sense and our tips, it's incredibly easy to lead a happy, fulfilled eBay life, where you can make decent amounts of money (more and more people these days are not only starting to make a significant second income out of the site, but are even quitting their day job and going into eBay selling full time!) and also pick up some brilliant bargains at the same time.

So we're going to cut through the nonsense and explain all you need to know to make your eBay time as successful as possible, in the firm knowledge that it should be fun as well! We're going to start off gently, with a whistlestop tour of the incredibly simple registration procedure, which lets you get up and running on the site in a matter of minutes.

→ Register with eBay

If you've yet to register with eBay – where have you been? It's not an exaggeration to say that the three minutes (literally) it can take to sign up and join the other ten million eBay users in the UK could be the most important minutes you ever spend online. Registration is a gateway to all the buying and selling services on eBay, the eBay community, its partners and all the worldwide sister sites. So, without further ado, lets get down to the nitty-gritty of signing up and joining all the fun.

Getting started

1 Although there are heck of a lot of things going on at eBay.co.uk's home page, everything's laid out pretty well so you can see just what you need to do. If this is your first trip to eBay, feel free to click on the eBay Explained link on the right-hand side to get a guided tour of just what you can expect on the site.

2 We're going to get registration out of the way first of all. Either click on Register in the very top menu, or click on the underlined link where it says Hello! Sign In or Register.

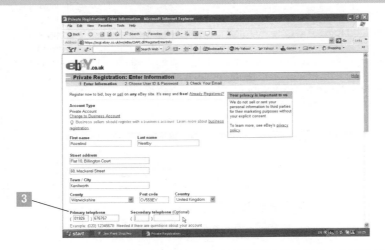

3 Personal detail time now. Work your way down the screen, entering your name, address, date of birth, phone number etc. All your information is totally secure.

For your information

Don't forget that your eBay username and registration get you into all the different eBays across the world – saving you loads of time and hassle when it comes to using any of the jam-packed sister sites for buying or selling.

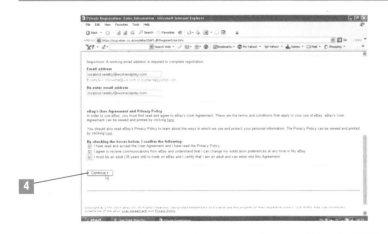

Did you know?

eBay has its own creed – a set of beliefs that it sees as crucial to the success of the site. Here it is:

'eBay is a community where we encourage open and honest communication between all of our members. We believe in the following five basic values:

We believe people are basically good.

We believe everyone has something to contribute.

We believe that an honest, open environment can bring out the best in people.

We recognise and respect everyone as a unique individual.

We encourage you to treat others the way that you want to be treated.

eBay is committed to these values. And we believe that our community members should also honour these values – whether buying, selling, or chatting'.

4

More details

4 You'll be asked to enter your email address – if you use a webmail address from the likes of Yahoo! or Hotmail, you may be asked later on to enter your credit-card details for added verification. If you use a work-based address then you're less likely to have to enter any financial details. Either way, your details are fully secure, so you don't need to worry. Click Continue after checking the legal boxes.

Important !

Don't panic if your credit or debit card details are requested during the registration process. The information here is strictly for identification purposes only, and absolutely no charge will be made to you whatsoever. Obviously, type your details in carefully to avoid problems. If you're really bothered, consider using a different email address – one from a company, school or paid Internet Service Provider should be satisfactory to bypass this step. As we said though, there is absolutely nothing to fear if you do find yourself having to enter financial details.

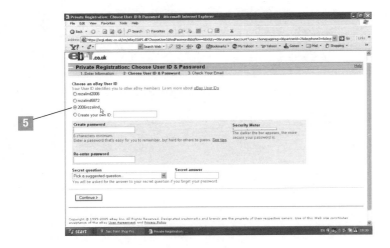

Password, please

5 Now it's time to choose a User ID. This identifies you to other eBay members. You can go for straight and factual, or something that describes what you see as your major eBay interest (such as 'filmfan2000' or 'sportsnut100', for example).

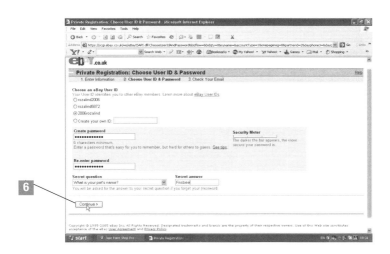

Password time

6 Then create your unique, easy-to-remember password, and a secret question and answer that will enable eBay to verify your identity if you forget your password and need a reminder. When happy, click Continue.

Timesaver tip

When you enter your password, you'll see what eBay describes as a Security Meter on the right-hand side of the screen. This handy device reinforces to you, in these days of scares about online security, how important good password practice is – in other words, make your password as secure as you can. Type in your password and the Security Meter bar will react to your suggestion – the darker the bar appears, the more secure your password is. You should be aiming to fill at least 2 notches, and preferably all, of the 3-bar meter to give yourself added peace of mind.

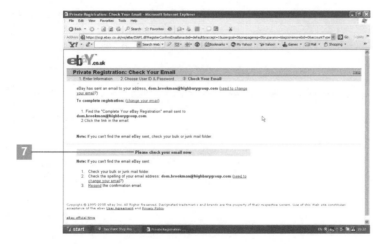

Email check

7 That done, and presuming you don't need to go through the extra step of adding the details of a credit card, eBay will now send you an email to the address which you specified in the signup. The email will have a link which you need to click on.

Important

Should your contact details change, it's a good idea to update them in the My eBay section of the site (you can see this link on the eBay home page). A well-updated personal home page will be looked upon favourably by people who are wondering whether to buy from you – we'll be looking at this aspect of eBay selling later on in the book.

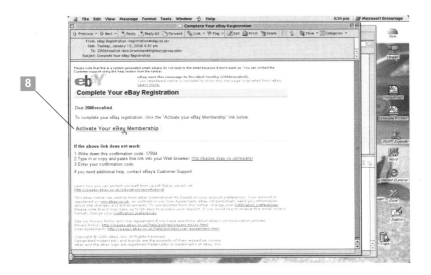

8 We open our email Inbox and have, indeed, been sent an email by eBay Registration. Simply click on the underlined Activate Your eBay Membership link in the body of the email. In the unlikely event that the link provided doesn't work, there's a backup procedure described in the email.

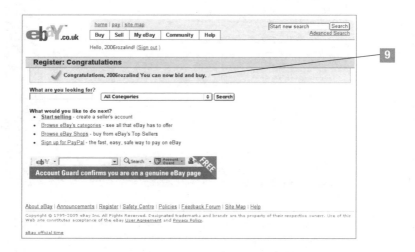

Important

Fake emails purporting to be from major companies and banks or eBay are the bane of many email users' lives. Click on these spoof emails and there's every chance you'll be suckered into a fake site where your details are taken and used for nefarious purposes. Worrying, huh? Well, thankfully, there are many ways to spot these kinds of spoof emails – we won't go into too much detail here, but a good clue is that the email will often be rife with spelling and grammatical errors – and there'll be a forged email address in the From line of the mail. To find out more about how to avoid these annoying and potentially damaging spoof emails, read eBay's very useful spoof email tutorial at http://pages.ebay.co.uk/education/spooftutorial.

Done and dusted!

9 You should now get a congratulatory message, all being well. As eBay says, you can now bid and buy! The whole process took about five minutes – the hand holding that eBay does here is most welcome, as it allows you to get down to the serious business of browsing eBay as quickly as possible.

Final thoughts

10 We've just got time to mention a couple of other pointers. During the
registration process, if you're concerned about privacy, click on the
underlined Privacy Policy link on the right-hand side of the first
Registration page, at the bottom of the other relevant pages, which will
take you to this screen. This should set your mind at rest.

11 More registration Help is on offer by simply clicking on Help at the top
of eBay's home page, then entering 'registration' in the Search Help
dialog box. You'll then be given further pointers to what is an easily
understandable process.

Of course, the history and genesis of eBay isn't massively important when it comes to actually buying and selling in the global marketplace, so we're not going to give a long-winded history lesson here. It's nice to get a bit of a background, however, as social history if nothing else (interested historians can get a more detailed company overview at http://pages.ebay.co.uk/aboutebay /thecompany/companyoverview.html). eBay was founded in 1995. Anecdotal evidence suggests that the springboard for its creation was the simple fact that French-Iranian computer programmer and internet enthusiast Pierre Omidyar wanted to help his wife indulge her hobby of collecting Pez sweet dispensers. Omidyar launched what was then known as AuctionWeb in September 1995, with the promise to visitors to provide 'an open market that encourages honest dealings'. The site grew, the name eBay and the logo came into creation, and the ball was well and truly rolling. We'll be dotting interesting eBay facts and figures throughout the book, but two figures stand out – eBay has a global customer base of 168 million, with 10 million in the UK alone. Other online companies would kill for such global reach, and the figures are going up even now. The eBay juggernaut, it seems, is more or less unstoppable.

→ eBay's partners

It's no surprise, of course, that such a global eCommerce giant as eBay has aligned itself with a good few partners. eBay's partnership with the payment monolith PayPal will be explored in more detail in chapter 5; its recent acquisition of Skype is described below. Elsewhere, interested surfers can click on the links at the bottom of the home page at www.ebay.co.uk to Shopping.com (http://uk.shopping.com/), a powerful shopping search engine which allows you to compare prices for millions of products across thousands of internet sites, to make sure you get the best deal on offer. Gumtree, meanwhile (www.gumtree.com), is an interesting online community, where you can find jobs, flatmates, friends and partners, and buy and sell goods, specific to your area – there are now Gumtree sites for 25 places in England, such as Leeds, Bournemouth and Reading, not to mention global presence in Australia, New Zealand, Poland, South Africa and beyond. The 20 new UK Gumtrees that have just been launched are an indication of the ambition of the site, so this is one eBay partner well worth looking out for. Keep a check on the eBay home page, as it would be no surprise if other major partners join the eBay family in the coming months.

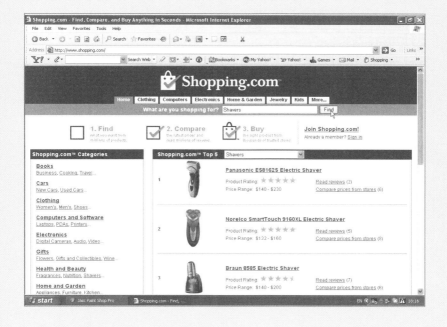

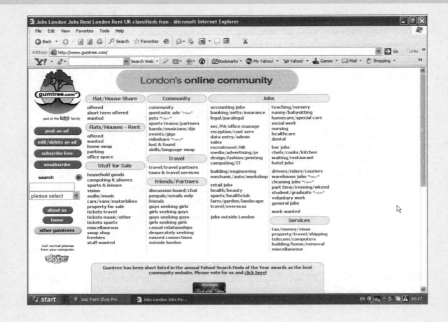

→ eBay around the world

When we're surfing the net, we get used to the fact that there are English and American versions of the major websites, such as Yahoo!, Amazon and Google. More and more of the really big players online, however, are branching out and understanding that different, localised versions of services can be a real money-spinner in the online world. eBay Ireland (http://www.ebay.ie/) launched last year, but that's just the tip of the iceberg; go on the UK home page and you can see links to (deep breath) eBay sites for Argentina, Austria, Belgium, Brazil, Canada, China, France, Germany, Hong Kong, India, Italy, Korea, Malaysia, Mexico, Netherlands, New Zealand, Philippines, Poland, Singapore, Spain, Sweden, Switzerland, Taiwan and the US. Your eBay username and password gets you into all of these country-specific eBays, so if you fancy a gander at what's on sale throughout the world, and your linguistic skills are up to it, it's well worth a whistle-stop tour around the globe. You could pick up a bargain – just be aware of the different rules that different eBays operate under.

→ eBay and Skype

When it comes to speculating about what the future holds for eBay, one of the most interesting pointers can be seen in eBay's recent acquisition of Skype, the world's leading internet telephone company, for a cool £1.4 billion. Skype (www.skype.co.uk) is a 'Voice Over Internet Protocol', or internet telephony, service – technical terms which basically amount to the fact that it's a piece of software which allows registered users to talk to each other via their PCs. Effectively, you're making phone calls via your computer, and saving a pretty packet at the same time. Skype has grown from humble beginnings to being incredibly popular with computer users across the globe, so the big question is… what does eBay want to do with it and its 54 million users? Well, it's not rocket science to speculate that Skype's incorporation into eBay will allow millions of auction users to chat with each other as a transaction takes place, sorting out all the small details that crop up when you're involved in an auction. The instantaneous, private conversation that Skype allows will potentially be a massive boon to eBay – one of the long-standing complaints about eBay has been the difficulty of sending countless emails to people when you're trying to sort out a deal, and this downer could now become a thing of the past. Keep your eyes peeled for exactly how this highly interesting deal pans out.

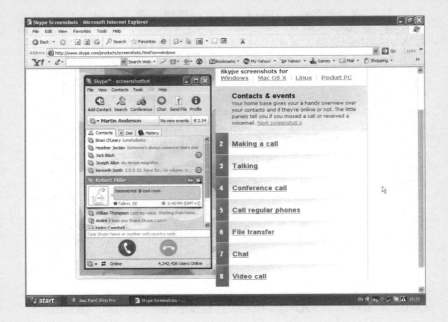

2 Buying on eBay

WHAT YOU'LL DO

- → Take a first browse through eBay
- → Carry out your first eBay search
- → Brush up on basic search tips
- → Learn about advanced searching
- → Track multiple items
- → Create a favourite search

- → Find and bid for an item on eBay
- → Cancel a bid
- → Find misspelled and no-bid listings
- → Use eBay buying resources
- → Discover some tools for eBay buyers
- → Learn some eBay security tips

As far as using eBay is concerned, the majority of your time is likely to be spent investigating the big two disciplines – buying and selling. Perhaps one of the best ways to approach the site, if you're a newcomer, is to get used to browsing and bidding for items for a couple of weeks, and then take a few dips into the world of selling once your confidence has built up.

So, in this chapter, we're going to look at different browsing techniques, how to carry out some stunningly effective searches, and how to get the most out of buying on eBay. The auction giant has a load of buying tools which will help you on your way – click on Buy from the eBay home page and then Buying Resources to find out just what's on offer. eBay has become so massive over the last couple of years, of course, that there may well be a few questions bugging you before you feel totally confident about joining the party. Questions about the reliability of a buyer, the legality of an item that's up for sale, what to do if things go wrong, and how to avoid fraud, are all, without doubt, relevant concerns. The good news, however, is that there are a lot of simple, common-sense tactics to protect yourself and avoid all those scare stories that national newspapers seem to delight in reporting. Checking a buyer's feedback, looking at his or her selling history, utilising the eBay Safety Centre and looking before you leap are all good tactics that will set you on the road to buying success, and we'll be mentioning all of these tips, plus many more, as we go through this chapter and beyond. And don't forget that eBay has numerous well-signposted help forums and resources to help you out if you get stuck.

So let's take our first steps on eBay with some simple browsing and item hunting.

If you're going to spend a lot of time on eBay, which will hopefully be the case, then you need to know clearly how it works from the very start: How things are arranged; what's on offer to you; where to go when things go wrong; where to go to chat to like-minded users; and how to get the very best eBay resources. So let's go on a whistle-stop tour of the main features of the site.

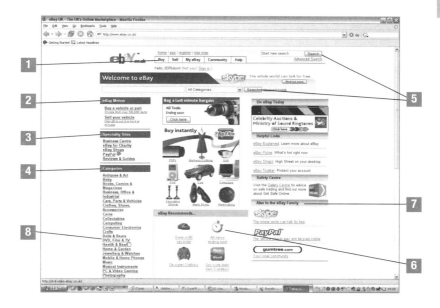

What's on offer?

1 Here we are on the home page of eBay.co.uk shortly after successfully completing our registration. There are many things to notice from this packed but well laid out home page, such as the main tabs at the top of the screen with which you can go to five of the main areas of eBay – Buy, Sell, My eBay, Community and Help.

2 Down the left-hand side of the screen are very important areas of the site. The growth of eBay Motors, where you can go for all your auto needs, has been such that it has its own separate area on the left, with links to help you choose from over 100,000 auto items, or sell your motor.

3 Then there are the speciality sites, which we'll look at in greater detail later in the book. eBay Shops, for example, is a huge area of eBay, where enterprising eBayers can collect all their wares together and sell under one virtual shopfront. The money you can make from this area of eBay is such that there have even been tales in the national newspapers about people giving up their daytime jobs to become full-time eBay shopholders, and earn thousands of pounds per week in the process.

4 The category list shows off some of the massive depth of eBay, helping you find everything from antiques to toys that much quicker.

5 Two search dialog boxes are accessible from the home page: in the top-left corner, and just under Welcome to eBay.

6 The centre of the screen is devoted to special deals, last-minute bargains, recommendations and any unique promotions that happen to be running at the time. Click on All Items Ending Now to find out which auctions are about to close – maybe you can sneak in and nab something from someone else's clutches!

7 eBay is a wide church, and on the right you can see some other members of the eBay family, including Skype and PayPal, which we'll return to later.

8 For the moment, lets have a browse through a random category – we click on DVD, Film and TV.

For your information

The eBay home page is an organic, ever-growing beast, so don't be surprised if layout changes appear from time to time. The basic principle of categories on the left, special deals in the middle and helpful links on the right is likely to remain, however.

For your information

We've pointed out the main features of the home page, but don't ignore what's right at the bottom – primarily, links to eBay sites around the world. Seeing links to eBay Argentina, eBay Philippines, eBay Singapore and the like reinforces just how massive eBay really is.

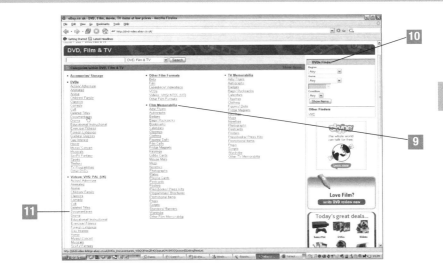

Category information

9 Click on any category in eBay and you'll most likely be taken to a screen similar to this one – where your main category is broken down into loads of different sub-sections. So here, for example, we have the sub-categories of DVDs, Videos, Other Film Formats, Film Memorabilia and TV Memorabilia.

10 There's even a handy DVDs Finder on the right-hand side where, in a matter of seconds, you can track down a DVD you'd like to find.

11 We've got a hankering for the documentary genre of DVDs, so we click on Documentaries to see what eBay has to offer.

Important

It probably won't surprise you to learn that the DVD category on eBay offers ample scope for people to try and flog items of questionable legality, as well as adult-only amateur home videos. A quick click on Other DVDs shows plenty of content on the rather seedier side of life, so just bear that in mind if you have any kids keen on browsing through the mountain of DVDs. eBay does have fairly strict guidelines regarding this kind of stuff, but there'll always be dodgy auctions that slip through the net, so take care.

What's on offer?

12 We get some 20,721 results for Documentaries – wow! That's 415 pages of results. With so many results, it's wise to study the listings carefully, as there's bound to be an anomaly or two – for instance, in this set of results, there's no way that the successful BBC2 sitcom *Coupling* classifies as a documentary. Down the left-hand side, you get further sub-sections of Documentaries, including News, Military, Crime and Science. These can help you narrow down your selection, or help if you're only interested in a certain genre of documentary.

13 As with any eBay browse, you can also specify a location to search through – you might want to only look for sellers within 50 miles of your own postcode, for example. You can set your choice here, as well as impose a whole other set of different criteria on the results that you get. For the moment we're only browsing though, so we don't need to worry too much about search variables yet.

14 We're big fans of documentary maker Michael Moore, so an auction of his film *Bowling for Columbine*, at a current price of just 99p plus £2 postage, seems very attractive. We click on the underlined link for the auction.

For your information

Don't underestimate the importance of a good, leisurely browse around eBay, especially when you're learning the ropes of the site. With just a casual browse, you can pick up what kind of items sell well on eBay, what kind of items are available, how to present your items effectively, and much more. Browsing and researching will also become crucial later on, if you want to become an effective eBay seller – there's no point, for example, in pricing yourself out of the market and offering the same kind of items for much more money than a rival seller. Get to know your market and you'll find your eBay life really starts to take off.

Timesaver tip

eBay also has a handy Compare function which we recommend you make the most of. If there's a couple of items that you're interested in – or maybe the same item on offer from two different sellers – then click the little tick box next to the one-line auction description, and then click Compare at the bottom of the screen. A new screen will appear, showing all the details of your chosen comparison at a glance, helping you make your mind up about which one is the right one to go for.

Look at the auction listing

15 With a few cosmetic exceptions, all eBay auctions look like this. The title of the auction is at the top of the screen, followed by important bid information such as the starting bid, the time left before the auction ends, the item location and the history of bids. (By the way, aesthetics shouldn't be underestimated when it comes to presenting your item in an auction. Grab someone's attention with clever use of fonts, headlines and descriptive text and you've won half the battle in persuading somebody to part with their hard-earned cash.)

16 On the right is the all-important seller information. Who is the seller? How many sales has he made on eBay? What's his feedback score? This particular seller has a mighty impressive positive feedback rating of 99.6 per cent. You can ask him a question, read everybody else's feedback comments, and view other items he has sold, all from this little dialog box. The feedback system is absolutely crucial to the ethos of eBay, and lets you make an informed judgement about whether you want to trust your money with the seller.

17 Under Description, you can find out loads more about the item in question – expect here to find a detailed rundown from the seller about exactly what's on offer, its condition, and whether there're any extra postage costs for you to factor in to the cost of the deal. Read this blurb carefully, and pay special attention to the payment methods accepted by the seller, as these can change from seller to seller.

18 And there you have a basic auction! Click Place Bid if you want to make a bid. Next, we'll dive into a very basic use of eBay's search engine, moving on to more advanced search options as we go along.

For your information

What categories are on eBay?

From the eBay home page, click on All Categories to see just what you get when you pay a visit to eBay. The major categories are: Antiques & Art, Baby, Books, Comics & Magazines, Business, Office & Industrial, Cars, Parts & Vehicles, Clothes, Shoes & Accessories, Coins, Collectables, Computing, Consumer Electronics, Crafts, Dolls & Bears, DVDs, Film & TV, Health & Beauty, Home & Garden, Jewellery & Watches, Mobile & Home Phones, Music, Musical Instruments, PC & Video Gaming, Photography, Pottery, Porcelain & Glass, Sporting Goods, Sports Memorabilia, Stamps, Tickets & Travel, Toys & Games, Wholesale & Job Lots, and, finally, Everything Else. Consider that each category has dozens of sub-categories, and that there're hundreds of thousands of auctions in virtually all of the categories, and you can begin to get a grasp on just how massive the eBay behemoth is.

Did you know?

Weird items on eBay

The world's press loves eBay, because virtually every day it throws up some fascinating human-interest stories, linked to the weird and wonderful things that people sell on eBay. Don't believe us? Over the last year or so, items on sale on eBay have included: the skull of a 200-year-old Hawaiian warrior (which landed the seller in hot water with the law), a cricket stump used in the Ashes series of 2005, Roy Keane dolls, a leaky tent (seller Iain McConachie attracted a massive 47,000 bids for this curio!), a hairbrush belonging to Angelina Jolie, a lunch date with Rupert Murdoch, bottles of air from Snowdonia and the Brecon Beacons (at £24 a pop!), the green Range Rover that Princess Diana and Prince Charles used for their first date, the Conservative Party (the listing was described by the seller as 'over 180 years old, rarely used in recent years and in need of major restoration'), Freddie Mercury's Volvo (allegedly), and numerous slices of toast... (it's best not to ask!). And that's just the tip of the iceberg, with more and more weird and wacky items hitting the eBay auctions every day. Keep your eyes peeled as you browse eBay or your daily paper, and you're bound to come across something that amuses, infuriates, baffles and astounds.

First impressions of the eBay home page can be a little misleading. With all the web links, category descriptions, special deals, recommendations and featured items, it can all seem a little bit disconcerting – how on earth are you going to get some efficient, effective searching going when there seems to be so much going on?

Luckily, the eBay search engine is an absolute doddle to use, and the good thing is that you can tailor your searches to be as wide-ranging or as specific as you like. In the same way that, hopefully by now, you've got used to getting the maximum from your Google searches, successful hunting through eBay relies on you having a clear idea of what you're looking for, and the flexibility to change things and enter different search terms if you're not getting what you want.

Here, we carry out a very basic search on eBay.

Start a search

1 There are obviously lots of different ways you can search on eBay, with two different search dialog boxes on this home page alone. For the moment, and for the purposes of showing just how powerful the eBay search engine is, we're going to use the basic search box in the top-right of the home page. We're on the look out for some Star Wars related merchandise for a nephew's birthday, but aren't quite sure what kind we want, so we just type in Star Wars and click Search.

Timesaver tip

Underneath Welcome to eBay on the home page, you can also carry
out a category-specific search. So if you know that you want to look
for a Liverpool Football Club DVD, for example, type in 'Liverpool' in
the search box, then use the pull-down menu to scroll down to DVD,
Film & TV. You should then be directed to a list of relevant Liverpool
DVDs for your viewing pleasure.

Results time

2 As expected, we get a stack load of different results. Just look at the
different categories that 'Star Wars' fits into – Toys & Games,
Collectables, Books, Comics & Magazines, Clothes, and Video Games,
to name just a few. Also, take a look for further info at the related
searches below the main box – 'lego' and 'light saber' may be
especially relevant as well.

3 We need to investigate Toys & Games, so under that category, we click
on Action Figures.

Timesaver tip

Get even more searching tips than are available in this chapter by clicking on the all-important Help button on eBay's home page. Then look under the Finding sub-section to get all sorts of searching advice, including how to manage your results and customise your results page.

4

5

More results

4 Here's the main screen from where you'll make all your decisions. On the left-hand side, you can see further search options, further matching categories, and a list of links to eBay shops which may have what you're after.

5 And in the middle of the screen you'll see a list of the most relevant auctions of the kind of things that you're after. Each one-line description has: a picture (if available); how many bids there have been for it; whether it's available as a Buy It Now item (where you skip the auction part of the proceedings, and just buy the item for a set fee instantly); postage costs; PayPal options (PayPal is the universally recognised safe way of paying for goods on eBay, which we'll explore later); and how long the auction has left. If an auction takes your eye, click on its underlined link to explore further.

There it is

6 Our search ends with a description of the auction. We'll go into more detail on the layout of this page later, but for now you can decide whether to bid for the item, ask the seller a question, or look at his feedback comments, which is what we'll do. And there we have a simple search – and rest assured, there are lots of ways to customise searching still further, which we'll explore next.

There are a number of simple tactics to employ that will help your eBay searching, without you even having to carry out an Advanced Search. Creating and then saving a 'Favourite Search', for example, is a good way of not having to enter the same search terms again and again; we'll take a look at how to do this later. The eBay toolbar, which you can download from http://pages.ebay.co.uk/ebay_toolbar/, can sit on your desktop and help you search for items in double-quick time, with just a quick click on the Search eBay button. Knowing exactly what you're looking for is a big help, of course. Type in 'Liverpool' for example and you'll get many, many different results in different categories. Type 'Liverpool FC Poster' into the search box and all of a sudden you've whittled things down to just 50 results.

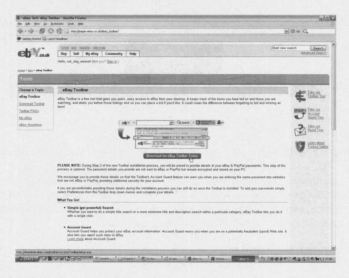

Remember that what you call something, a seller may have another term for – are you looking for a 'sofa' or a 'couch'? Try both terms to get the kind of results you want. And try different spellings of terms, as some people's English may not be as perfect as your own! Really sure about what you want? Then be specific! Type in 'Dolce and Gabbana Blue Dress' if that's what you're after, rather than just 'dress' or 'blue dress'. You'll narrow down your results incredibly. Don't bother with the words 'And' 'Or' and 'The', as those words are included in a search anyway.

These are just a few tips, and they obviously depend on the kind of searches you're running. As on some occasions you may actually want to increase your search results and have hundreds of options, rather than narrowing your search down to just a few. It all depends on the individual search. If you take some of these tips on board, however, along with the principles of advanced searching, you can't go far wrong.

So we've had a look at getting the most out of browsing and some basic eBay searching – it's now time to try an Advanced Search. Don't let the term 'Advanced' put you off. Basically, what we're going to be doing is applying different sets of criteria to our results, based on such variables as the location of the seller, how soon the auction is ending, and how much the item is priced at. There are a fair few variables to get your head around here, so let's get on with some advanced searches without further ado.

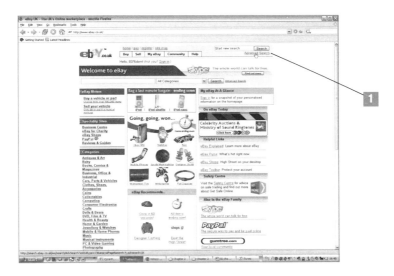

Carry out an advanced search

1 An Advanced Search puts the power firmly in your hands, letting you specify as many different requirements for your auction results as you want. Let's put the system through its paces. From eBay's home page, click on Advanced Search.

2 It's best to take an advanced search in different stages, as there are so many different variables for you to play around with. We're going to look for a new album by a singer called Beth Orton, entitled *Comfort of Strangers*. For the moment, we just enter that title into the search box, then choose the Music category and click Search.

Timesaver tip

If you know of a particular seller – maybe you've bought something from him or her before, or you've seen good things about them in your eBay browsing – you can opt to search by seller. Click on Items by Seller in the search box on the left of the Advanced Search screen, then enter the seller's user ID to see items just from that seller.

Timesaver tip

You can also search for items in eBay shops through the Shops option in this search box. Click Find Shops to unearth a shop that might sell your chosen item, or click Items in Shops to do a search through lots of different shops.

Results

3 That gives us 50 results, which isn't bad, and gives us something to go on. There are plenty more things we can do to enhance our search, however, so we click on Advanced Search at the top of the screen once more.

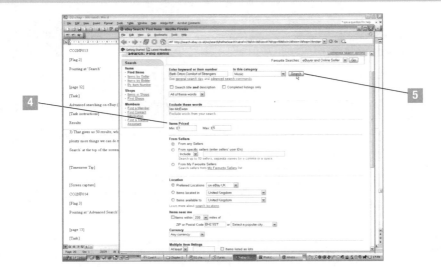

4 Time to get clever. We know there's a book by Ian McEwan with the same title as our CD, so we opt to exclude 'Ian McEwan' from any of our results. We then specify a price range of between 1 and 5 pounds – we don't want to see any results that fall outside this bracket.

5 We then click 'Search' again.

Timesaver tip

eBay lists a series of different search commands that you can use in your searches to get highly relevant results. It's a similar process to how you would use search commands in Google, for example. Put speech marks around a search term to get results in a particular word order only, e.g. 'Birmingham City'. If you're looking for a red shirt that isn't made of silk, type in 'shirt red – silk' to exclude any silk results. The minus sign here is the all-important string which tells the search engine which word to exclude. There are many more of these kinds of tactics, and to get the full list, take a look at http://pages.ebay.co.uk/help/find/search_commands.html?fromFeature=Advanced%20Search.

More paring down

6 We're now down to 16 results. Let's keep going. Click on Advanced Search again.

7 Now to set even more criteria. We change the price range to 1 and 3 pounds only, demand that the results shown are UK only, within 100 miles of our postal area, open to PayPal users, and in a listing that ends within the next 2 days. That should sort things out a bit!

8 It certainly does – we're down to just 2 results now. Fewer search options gives more results, as eBay kindly points out in a box below the auctions, but we can now use the handy Compare option to see which one of the 2 auctions floats our boat the most.

Timesaver tip

If you're looking for a particular member on eBay, you can do a member search under Find a Member in the Search box. Just enter their user ID or email address to find them. If you enter the email address only, you'll just get limited information back, for security reasons.

Timesaver tip

Click Customise Search Options from the Advanced Search screen to decide exactly what search options you want eBay to provide you with.

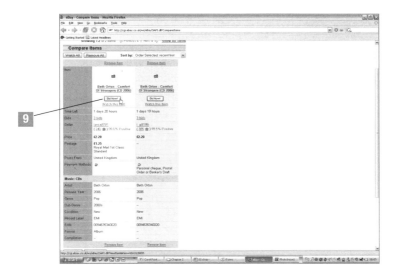

Let's bid

9 It's a close run thing, but we decide to go with the auction on the left of the comparison divide. We now know that the auction satisfies a significant load of different requirements, which gives us the confidence to bid and know that we're going to get exactly what we want. Click Bid Now! to bid.

For your information

Advanced searching tips

So, we've uncovered quite a few ways of getting the results you want from eBay, using the price range, geographical location and payment method, to name just three elements, to whittle down what eBay's engine gives you. You can also use searching as an invaluable research tool, of course. If you're planning on selling a Harry Potter book, but don't know how to price it, click in the box next to Completed Listings Only in an advanced search. This yields results only from auctions that have closed. That way you can make sure, at least to a certain extent, that you don't overcharge your punters. Remember not to get disillusioned too easily with searching. If you demand too many requirements, then you won't get any results back at all unless you're really lucky. So you may have to widen your net a little bit, and not be too inflexible, to adapt to the eBay market. If all your searches prove fruitless, you can even post an announcement through eBay's Want it Now facility at http://pages.ebay.co.uk/wantitnow/index.html, and let millions of people know what you want. With a bit of luck, someone will be reading your post and be able to help you out pronto!

→ Track multiple items

One good way of keeping your eye on your searches in eBay is to 'watch' them through a useful eBay feature. After all, you're not going to want to bid straight away on auctions that catch your eye – you might want to bide your time and see how the bidding goes before deciding to commit to a financial bid. Browse for an auction as usual, and when you find one that catches your eye, click on the Watch This Item in My eBay link from the main auction listing screen. My eBay is a useful part of the eBay experience that has grown in functionality over the last couple of years, to the stage where it can become a really useful 'base' for your eBay activities. There's a My eBay tab on the home page of eBay – click on it to see all sorts of information about your eBay life – including which items you've designated to 'watch'. You can then see at a glance the state of the auction and how long it has to go – you can also opt to get an email from eBay informing you of any activity in your chosen auction. And then, if you choose to, you can obviously bid quickly for an item. It's probably best not to track too many items at the same time, as it could get a tad confusing, but certainly feel free to have a few on the go at the same time. There's plenty more you can do in 'My eBay' as well, which we'll explain later.

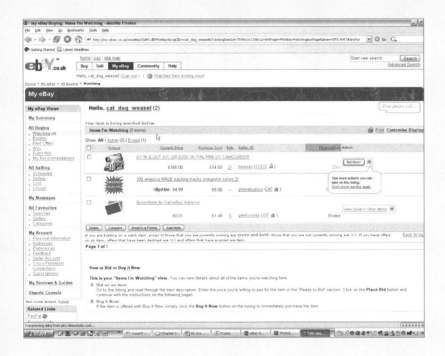

→ Create a favourite search

If you carry out the same kind of search frequently on eBay, you can save a good deal of time and hassle by creating a Favourite Search. This lets you save the searches you run, and get any relevant new results emailed to you, keeping you on the ball and making sure you don't miss something important. Let's take a look at how Favourite Searches work below – for more info, try http://pages.ebay.co.uk/help/buy/search_favorites.html.

Starting off

1 Begin proceedings as you would do normally, with a standard search. We're going to search for an online-only show by Ricky Gervais, one of a series of weekly 'podcasts' from the comedian. The fact that he releases a new half-hour's worth of material every week gives us an idea that this might be a search which we'd like to get eBay to keep on top of every day for us. Click Search as normal.

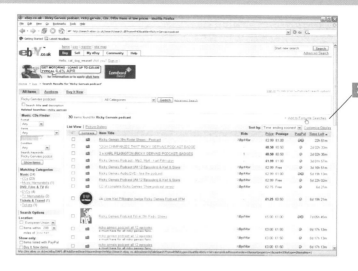

2 The results come back – 30 of them in this case. As we've mentioned, this is a search which we reckon we're going to be carrying out pretty often for the next few weeks, so we click on the handy Add to Favourite Searches link in the top-right corner.

For your information

eBay allows you to have up to 100 Favourite Searches, which should be ample for your requirements. You probably won't have the time or the energy to have that many searches, and it probably isn't that good an idea anyway, as all those results will clog up your Inbox and make life needlessly confusing. It's probably best just to stick to a handful of Favourite Searches unless you have a powerful memory and loads of patience to plough through dozens of emails every day.

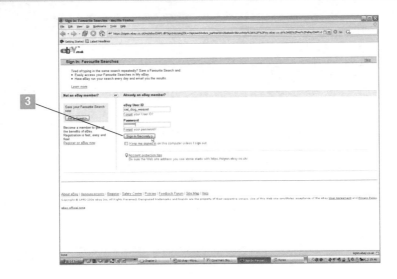

Log in and set parameters

3 You might be asked to re-enter your login details, just for security purposes. Type your user ID and password, then click Sign In Securely.

Timesaver tip

Deleting a Favourite Search when you're tired of it is a doddle. In My eBay, tick the little box next to your designated search, and then click Delete. Then it'll be gone.

Timesaver tip

You can quickly add a new Favourite Search from within My eBay by simply clicking on the Add New Search link.

4 You now get the option to add your search to your list of Favourite Searches. Under Save Options, choose Create a New Search.

5 Under Search Name, make sure your search has a name which you'll easily remember.

6 Do you want eBay to email you with new items in your search field? You do? Tick the box next to Email Me Daily.

7 Then choose for how long you want to receive the emails – we select a period of 1 month.

8 Once you're happy with the choices that you've made, click Save Search.

Did you know?

eBay runs over 1 million searches for its members every day. That's a lot of donkey work!

Back at my eBay

9 Here we are back at the all-important My eBay section of the site. Now, under My Favourite Searches, you can see all your saved searches, and how many days left there are in each. If after a while you decide that you want to change things – maybe you've bought an item from the search and no longer wish to be notified about it – click on Edit Preferences in the Action column on the right.

For your information

The All Favourites part of eBay not only allows you to store your Favourite Searches – you can also make notes about your favourite sellers if you regularly deal with sellers who offer you excellent items and customer service, and even your favourite categories. This latter option would be especially useful if you're interested in a rare or unusual category, as with My eBay you would be able to go straight to it without having to hunt around for it in the normal full category lists.

Now that we've sussed out browsing and auction search techniques, we're going to go through some of the specifics to do with buying an item on eBay. Certain procedures will crop up again and again, but we make no apology for repeating key snippets of advice, such as always keeping a very close eye on a seller's feedback, and making sure that the post-sale correspondence is handled in the correct manner. We're going to teach you to possibly get one step ahead of the crowd with 'sniping' programs – utilities that you can download from the internet which bid for you during the very last seconds of an auction, theoretically increasing your chances of winning. Sniping is frowned upon by some eBay purists, but we reckon it's just part of the fun and the general process of bidding online, so don't let other people put you off. Indeed, the only real downer with sniping is that it has become so popular that in any one auction, there could be multiple people using these kinds of programs at the same time. Which makes for some very frenetic last-minute bidding!

Anyway, let's find an item to buy, and look at all the different ways you can approach snapping up that must-have item.

Starting off

1 At eBay's home page, www.ebay.co.uk, click on Buy.

2 We know exactly what we want – an Xbox 360, one of the hottest games consoles at the moment. We type it into the Search box and click Search.

3 We've got a few too many results to deal with comfortably, so we opt to narrow down the 3,735 matches by modifying the search to 'Xbox 360 Consoles'. We're interested in the machine itself, not any of the games or anything else with '360' in the title.

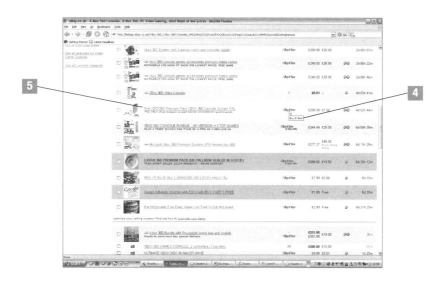

Results time

4 That's better – we now have much more specific results. Do you want to try and buy the item through an auction, or through Buy it Now? Buy it Now lets you get your hands on an item at a set price, without the hassle and unpredictability of bidding for it.

5 Click on the auction's title to progress. We're going to look at the Buy it Now procedure first.

6 The details of the auction appear. Click on Buy it Now.

For your information

If you're feeling benevolent, you can also check out the eBay for Charities section of the site, where you can buy a hot item and also do your bit for good causes at the same time. Look out for the blue and yellow ribbon on an auction as an indication that it's a charity lot – buyers can find out more at http://pages.ebay.co.uk/community/charity/buyerinfo.html.

Buy it Now

7 The Buy it Now price and quantity is confirmed. Click Continue.

8 Review the item. Postage and packing has now been added to the total price. Commit fully to buying by clicking Commit to Buy. Buy it Now is pretty simple, isn't it? That's why people love it. It's especially handy, obviously, if you're pressed for time and absolutely certain that you want to buy the item.

9 Now let's look at buying tactics in an auction where Buy it Now isn't available. Click the title of the auction which interests you.

Timesaver tip

It's worth repeating the general point that you really should feel totally at liberty to email the seller with any question you have about a listing, within reason of course (no one wants to be bombarded with dozens of pointless, irrelevant questions). Seeing how a seller deals with an intelligent question is a vital clue to whether they're someone with whom you should be doing business. If they seem evasive, shifty, or unable to answer a valid query, then you should really think about taking your business elsewhere.

Feedback

10 No apologies for diving into looking at a seller's feedback ratings again, as it's so crucial. After clicking Read Feedback Comments under Seller Information on the main auction page, you can further narrow things down by using the pull-down menu on the right to view feedback during a specific time period. We go for Past 12 Months and click Go.

11 Spend time reading the feedback. This guy has 100 per cent positive feedback and a lot of happy buyers.

12 Back at the auction, we have a question we'd like to put to the seller, which will help us decide whether the item is for us. Click Ask Seller a Question.

Timesaver tip

Don't be afraid to trawl your net far and wide when searching for that elusive item – and remember that eBay truly is a worldwide concern. eBay's Global Trade home page at http://pages.ebay.co.uk/ globaltrade/ shows you all of eBay's worldwide sites, which you can go to in an effort to get just what you want. The international discussion board will let you get answers to key questions that arise when buying from abroad – topics that will crop up more in this area of eBay life include international shipments, converting currency and even dealing with language difficulties. Make sure you're fully aware of all these issues before buying globally – there are plenty of help links on the Global Trade home page to help you fully understand what it's all about.

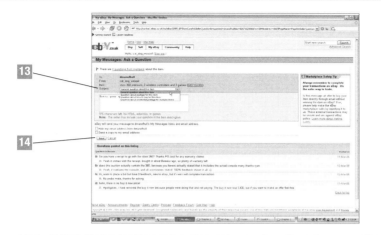

Question time

13 On the screen that appears, use the pull-down menu to choose a subject title for your question. Useful for the seller if he's getting lots of queries.

14 Then write your question. Keep it brief and to the point, then click Send to ask away.

15 Whilst you're waiting for the item, you can choose to watch the item and monitor its progress, by clicking Watch This Item in My eBay on the main auction page.

16 The link then changes to inform you that you are indeed watching the auction's progress, along with 2 other items. Click 3 Items to view your watched listings on one screen.

Timesaver tip

Be very suspicious of a seller who hasn't bothered to supply decent photographs for his listing, as this could show he's got something to hide. The availability of quality digital cameras these days is such that if a seller can't be bothered to get a decent photo online, they're probably not the thorough, conscientious person you want to be dealing with. Conversely, if someone's gone to the trouble of photographing items from different angles, especially on items such as furniture or electronics where it can be difficult sometimes to judge questions of scale, then they should go up in your estimation.

For your information

Check out the excellent Know Your Seller link at http://
pages.ebay.co.uk/help/confidence/know-seller-ov.html to get some
inside information on whether a seller is worth dealing with.

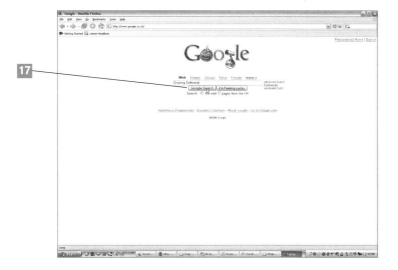

17

A diversion

17 Now for a diversion. We're going to show you how to use sniping
software – software which automatically bids for you in an auction, in
the last seconds (if you ask it to), to increase your chances of winning.
We'll go the long way around and show you how to find a program and
download it – obviously you'll only need to go through the download
process once, as the utility will sit on your desktop after that. If you
don't know where to find sniping software instantly, use good old
Google (www.google.co.uk) to help you out. Type 'Sniping software'
into the search bar and click Google search.

For your information

One of the most important things you'll be looking for in your eBay travels is item condition – listings that specify exactly what the state of the item is will help your buying confidence. Sellers have a Condition drop-down menu when they create their listing, and when you're browsing through listings as a buyer, you can tick the Show Only Items in New Condition box to get condition-specific results.

Find a suitable program

18 You'll see a good few useful links. Which one you choose is up to you and a matter of personal taste, really – maybe ask any eBay friends which ones they like. We go for the software from www.auction-sentry.com.

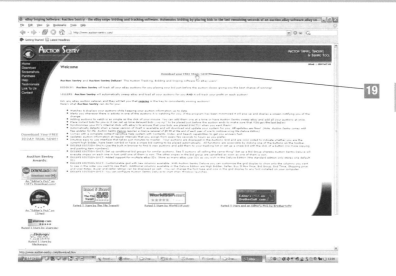

19 You'll be taken to the program's home page, where you can see the download link at the top. Click on Download Your Free Trial Now.

Timesaver tip

Another way of nipping into auctions at the last minute, is to use the Going, Going, Won… box which is to be seen on most general buying pages. You can then view nearly finished auctions for items such as watches, mobile phones, widescreen TVs, toys and much more, and quickly make a bid if you see something that you like.

Download

20 Confirm the download again on the next screen.

21 Instruct your Internet browser where you want the executable file to be saved to, and then click OK.

Timesaver tip

Don't forget eBay Shops are a very good place to make a buying decision. We'll be looking at both browsing through eBay Shops (http://stores.ebay.co.uk/) and building your own store in chapter 4.

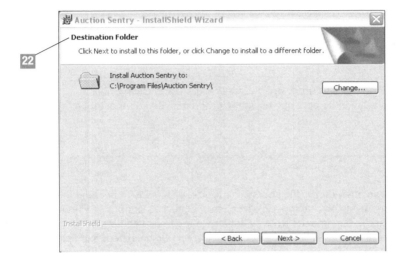

A wizard time

22 With the program downloaded, click on its icon, which should be on your desktop, to set up the InstallShieldWizard. This will set the program up for you properly, in a matter of seconds. First, decide which desktop folder you want it to be installed in. Click Next to go through the steps of the wizard.

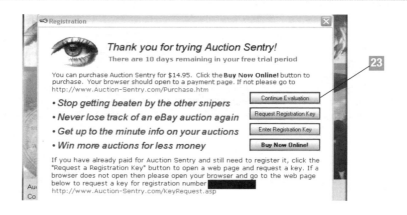

23   Once that's all done, open up the program and click 'Continue Evaluation' to continue using the program in its free-trial mode.

Timesaver tip

Every so often during the course of a typical year, an item becomes such a 'must-have' that people's critical faculties go out of the window, and there's a stampede to eBay to pick up the treasured item at whatever massively inflated price it's going for. Typical recent examples include the Xbox 360 in the UK – stocks at Christmas 2005 in the high street were criminally low, so people flooded eBay looking for the hot console. Sellers picked up on the high demand, and the number of Xbox 360s which purported to be official UK consoles, and which were actually nothing of the sort, was staggering. People's panic, especially just before Christmas, led them to bid desperately for the item without first checking out its specifications – and if your brand new £400 games machine doesn't work in the UK due to regional-specifications not made clear in a listing, you'll be pretty sick. Don't let the desirability of an item cloud your judgement and ability to check that you know EXACTLY what you're getting.

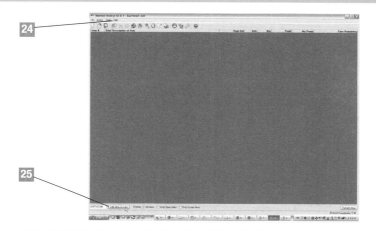

Use it

24 You'd be well advised to use the superb Help files first off, to get a full idea about how everything works.

25 Once you've digested how the program bids for you, enter the number of the auction(s) you want to track at the bottom of the main screen. Click Add eBay Auction.

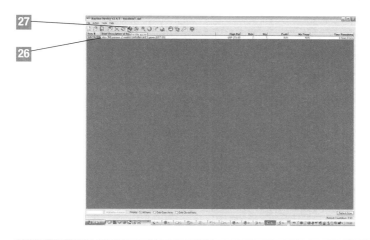

26 The auction appears at the top, with some of its specifics. Obviously you can have loads of different auctions being tracked in this window.

27 Click Bid On This Auction to instruct the program how you want it to bid for you, and when.

Timesaver tip

Why not shop by looking at pictures first so you can get a very quick visual indication of what's for sale? eBay Gallery (http://pages.ebay. co.uk/gallery-index.html) lets you shop by pictures. This can be an excellent way of sorting out in your mind what you want to bid for.

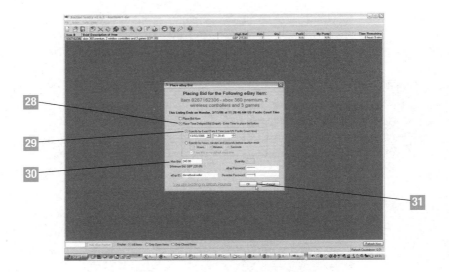

Set your commands

28 Specify what you want the program to do. We want a Time Delayed Bid – i.e. a snipe.

29 Set the exact date and time you want the snipe.

30 And the maximum bid.

31 Confirm your eBay ID and password, then click OK.

32   Then let the program do its work! As a final alternative, we'll just confirm how to place a normal bid. Back at the auction, click Place Bid.

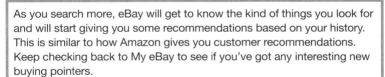

For your information

As you search more, eBay will get to know the kind of things you look for and will start giving you some recommendations based on your history. This is similar to how Amazon gives you customer recommendations. Keep checking back to My eBay to see if you've got any interesting new buying pointers.

Specifics

33 State your maximum bid for the item. eBay will automatically bid on your behalf until this maximum is reached (when it will inform you and ask you how you want to proceed). Click Continue.

34 Review the bid before agreeing to a legally binding contract to purchase the item if you're the winning bidder by clicking Confirm Bid.

Timesaver tip

Don't forget the Everything Else category on eBay, once you've checked out all the main categories. It's just possible that what you want is nestling in here – there really is a cornucopia of things in here, although beware of the countless rubbish 'get rich quick' schemes that are floating around.

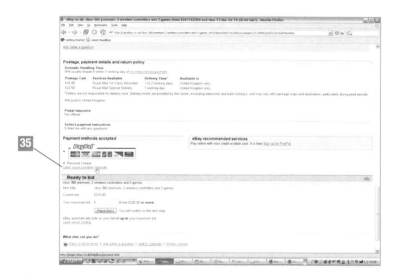

Payment

35 For each auction item, you need to be clear about how you're going to be allowed to pay for it, should you win. Click Learn About Payment Methods in an auction listing to get the full lowdown.

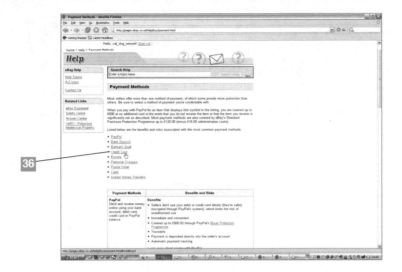

36 All the different options are explained in eBay's Help page on the topic – we really do recommend PayPal when it comes to safe and sound online payment. In chapter 5 you can learn about how to sign up for the service in minutes. Click any other method, such as Credit Card to find out more info about how it works.

Timesaver tip

Get the very latest eBay news from the general announcement board, at http://www2.ebay.com/aw/marketing-uk.shtml. You may be able to pick up valuable info – maybe advance warning of a '50 per cent off sale' day, for example, or just technical information about what's happening to different parts of the eBay universe.

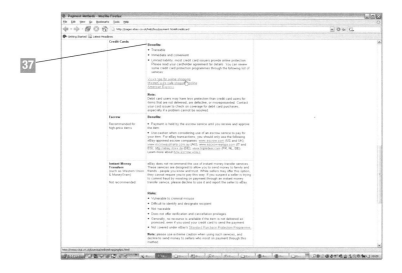

Pros and cons

37 The benefits of paying via each method are shown to you – paying by credit card does have a couple of benefits, such as being immediate and convenient, as detailed here.

38 If you do win the item, how things work after the lot has closed is almost as important as the bidding procedure itself. What happens will vary from sale to sale, of course – you may need to contact the seller a couple of times by email to thrash out any outstanding issues such as delivery times and payment procedures. On the other hand you may be lucky and have a seller who's using the eBay Checkout service, which limits the amount of time-consuming emails you need to send the seller, by calculating the total amount due and sending your payment to the seller. More info about what to do after winning an item can be found at http://pages.ebay.co.uk/help/buy/winning-ov.html, where we click Checkout to find out more.

Timesaver tip

If during your eBay searches, you're struck by the absence of a category which you think would enhance the site, why not make a suggestion in the Community area? There's a Category Chat board where you can pitch your case to eBay staff – you never know, you could just help to change the look of the site for ever.

Checkout tutorial

39 To find out Checkout's benefits click on Checkout Tutorial for Buyers. If you're convinced by it, you can add it to your own eBay sales if and when you decide to sell online.

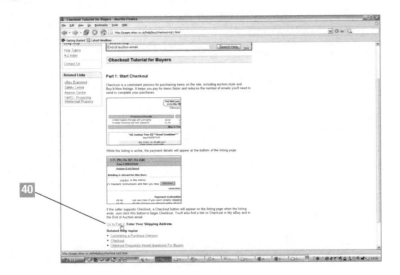

40 A step-by-step tutorial begins – read the info and click the link at the bottom to go to the next page.

Timesaver tip

We can't over emphasise the importance of taking time out to write a few lines of feedback for your seller, once a deal has been completed and you've received your item. Feedback is ultimately the key principle on which the whole of eBay works, so if you've been treated well, let the world know!

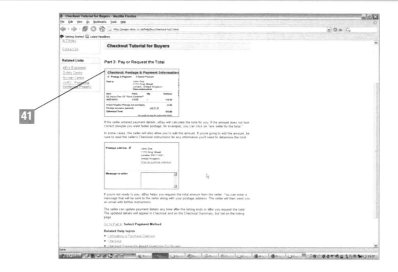

That's it!

41 Confirmation of how the Checkout procedure calculates the total costs is next. Getting this information instantly really is a boon.

42 And it's all wrapped up in a handy Checkout summary, which you can access from My eBay. Don't panic if you can't go through the checkout procedure – a couple of well-chosen emails should be enough to get everything sorted, if you're dealing with a decent and experienced seller. Don't be too impatient, and don't be impolite, as this part of the procedure is being rated by both sides – you could get bad feedback as a buyer if you're unreasonable. If your seller does let you down, and you don't receive your item, or get something that was different than described, pop along to the safety centre (http://pages.ebay.co.uk/safetycentre/) to detail your problems. Hopefully everything will go swimmingly, and you can gain experience as an eBay bidder, learning how to play the markets and pick up great bargains.

For your information

Other examples of sniping software

Unsurprisingly, when you learn of the popularity of sniping programs, there are plenty of other rivals to Auction Sentry on the market. We're big fans of Onbidder (www.onbidder.com), from the same people who brought you the internet service Onspeed, and you can also take a look at PowerSnipe (www.powersnipe.com/), Auction Stealer UK (www.auctionstealer.co.uk), Bidnapper (www.bidnapper.com), and many, many more. Browse through a Google search and see which one takes your fancy.

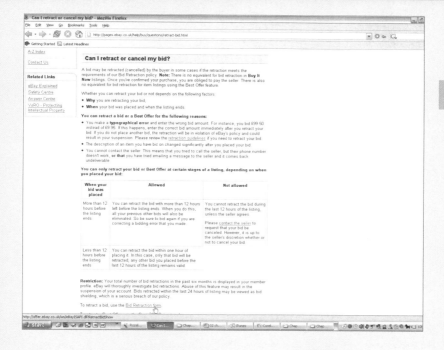

If you've made a typographical error (bidding £95 instead of £9.50, for example), or the description of the item changes after your bid, or the seller is uncontactable, you have grounds to retract a bid. There are restrictions, however, and eBay is likely to investigate retractions you make to check that you're complying with site policy. If you genuinely think you have a case, read the terms and conditions and use the Bid Retraction form that you can find at http://pages.ebay.co.uk/help/buy/questions/retract-bid.html.

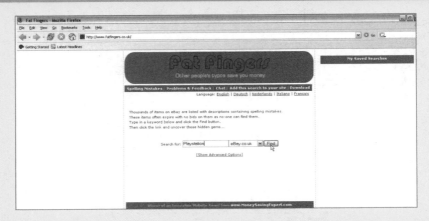

Many items on eBay attract no bids, not because they're sub-standard or inferior, but due to an unfortunate spelling mistake on behalf of the seller, which leads to it being hidden away from the main listings for an item. Spelling mistakes don't reflect massively well on the seller, but if you're willing to give them the benefit of the doubt – and we all makes mistakes, after all – why not try the useful site at www.fatfingers.co.uk. Type in your search term here, then click Find, and you'll be taken to eBay listings for typographical variations on your term. So typing in 'Playstation' brings us to listings for all sorts of dodgily spelt listings, including 'Palystation', 'Platstation', 'Playdtation' and 'Playstatoon'. You may well find a bargain here, hidden away due to over-eager typing hands.

For your information

What to look for in online auctions

For a summary of some of the buying tips we've been going on about here, try eBay's Buyer Checklist, at http://pages.ebay.co.uk/ help/buy/buyer_checklist.html. There's nothing too controversial here, but there's the odd extra useful pointer, such as having the foresight to check that the item listed is actually allowed on eBay (check the policies at http://pages.ebay.co.uk/help/sell/ questions/prohibited-items.html).

When you start to buy on eBay, or any of the hundreds of thousands of linked auctions that are currently active, you can access a host of useful Buying Resources from a box on the right-hand side. Full tours around eBay, a handy new Reviews and Guides section, the Safety Centre, PayPal tips and the chance to download the eBay Toolbar are just some of the goodies on offer here. Click See All Buying Resources and then have a browse around…

Buying Resources

1 eBay's Buying Resources help you get the most out of buying from the site. When you click on Buy, on the site's home page, the list of Resources is on the right-hand side. We'll look at a couple of the features here. We've already mentioned the eBay Toolbar, so let's click on the new Reviews & Guides feature.

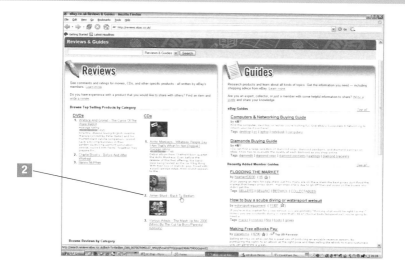

2 Reviews are specifically about the product, rather than the seller – still useful, if you're not sure about how good a CD, DVD or game actually is! We click to see what people are saying about the chart-topping CD from James Blunt.

For your information

We love the eBay Explained area of Buying Resources (http://pages. ebay.co.uk/help/ebayexplained/index.html) – in the same way that a search on Google can send you off down dozens of unexpected roads, just browsing through this part of the site is bound to tell you something you didn't know, no matter how many books or magazines you've read. eBay excels in giving you snippets of useful advice backed up by plenty of screenshots (a bit like this book, in fact!) – it doesn't patronise or waffle on in jargon-heavy language, either. The writers of this area of the site deserve a good deal of credit.

Read and digest

3 There are lots of positive comments about the album, which may just convince you to shell out if you were unsure before. Click Yes next to Was This Review Helpful if you found any particular words of wisdom informative and handy.

4 If your mind has been made up, click Buy One Like This to see the latest James Blunt offers.

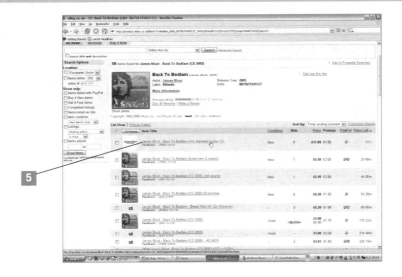

Buy it

5 Then just click on an auction title to proceed as normal with a bid.

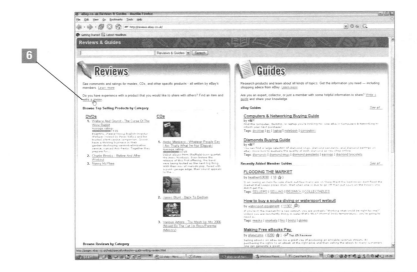

6 Back at the reviews home page, why not click on Write a Review and add your own comments about something?

Important

Remember, what looks like an incredible Buy it Now price on eBay may not be quite as good once you've factored in postage and packing charges. These are not included in the initial Buy it Now price you see in a listing.

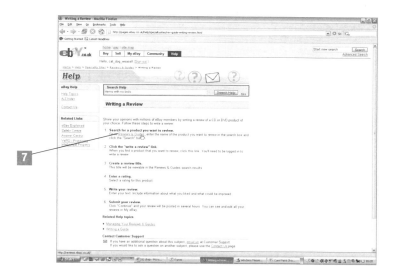

2

Understand it all

7 You'll be taken to the Help pages to ensure that you know exactly what writing a review entails. Click Reviews & Guides when you've read the blurb.

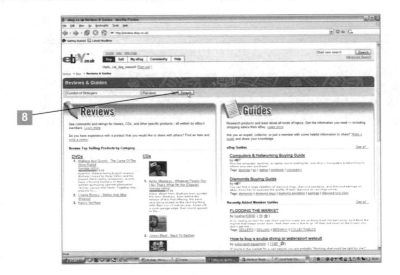

8 Type in the title of the product you want to review, then click Search.

Timesaver tip

In the unlikely event that there appears to be activity in your eBay account not instigated by you – maybe your password seems to have been changed, or your username is being fraudulently used to list items that have nothing to do with you – go to the Safety Centre and click the circle next to Suspicious Account Activity. The root of the problem should then be able to be uncovered pretty quickly.

Be the first!

9 Three results were found, of which two are relevant. If no one else has reviewed it, as is the case here, click Be The First to Review It!

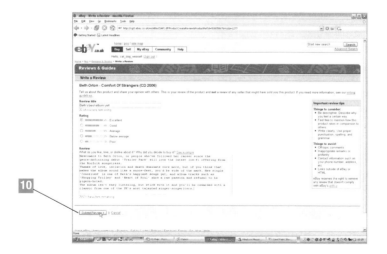

10 Then write your review, and click Submit Review. Ultimately these reviews could really prove to be a boon when you're trying to make up your mind about an item – with the number of people that use eBay every day, just think about what a huge database could build up.

For your information

The Reviews part of the Reviews and Guides area does seem pretty shallow at the moment – we guess that's due to the fact that it's a pretty new feature on eBay. Hopefully as time goes by, more and more people will feel encouraged to write their product reviews, and help to build up an invaluable knowledge database for buyers.

Moving on

11 Well done, your review is finished. Click to go back to the Reviews & Guides home page.

12 Guides, meanwhile, on the right, can give you invaluable advice about particular areas of eBay life. We click Diamonds Buying Guide.

Timesaver tip

In the same way that you can write your own review of an item, you can also write your own guide to aid other buyers. eBay's own tips for this process can be found at http://pages.ebay.co.uk/help/specialtysites/rev-guide-writing-guide.html; valuable points to remember include obvious ones like, not writing anything libellous or profane, not giving out your personal information, and not linking to other sites outside of eBay.

Diamonds are forever

13 Brilliant, in-depth advice about this category on eBay, and what to look for when you're buying, appears. It's extras such as this which really add to the eBay experience.

14 Use the Reviews and Guides whenever you want, as a valuable background buying tool – just what Buying Resources are all about. Let's take a quick peek at the Safety Centre next, clicking the link from the main Buying Resources box on the right.

15

Help, please

15 Buying and paying safely is crucial on eBay, and the Safety Centre can direct you to loads of links that ensure you understand fully the principles of security.

16

16 When you click on the link you'll get a whole host of topics to investigate further, including payment protection, a buyer's checklist and how PayPal works.

For your information

eBay's Help pages, which you're likely to need at least a few times in your eBay life, even have the facility now for you to rate how helpful they've been! A box on the left-hand side of a specific help page now lets you answer Yes or No to Was This Page Helpful, and then gives the option to leave comments about how to improve the page if your own experience makes you think that some other advice could have been included.

Protection

17 Useful tips on how to protect your eBay Account can be accessed from the area on the far right, next to the padlock.

18 Don't forget the eBay toolbar, accessible from a link in the Safety Centre. As we've mentioned already, the convenience the utility offers, plus its Account Guard facility (helping protect you from spoof websites) make it well worth a quick free download.

For your information

If you're lucky, you may see a listing with a Get it Fast logo on it, which means that the seller can ship your item within one business day of receiving your cleared payment for the item. Take a look at http:// pages.ebay.co.uk/help/buy/get-it-fast.html for the lowdown on how Get it Fast items differ from the norm.

Stop the spoofs

19 Spoof emails are getting more and more prevalent these days, sadly. Learn how to spot one, and report it to eBay, by clicking Spoof & Phishing Advice in the Safety Centre, and being taken here.

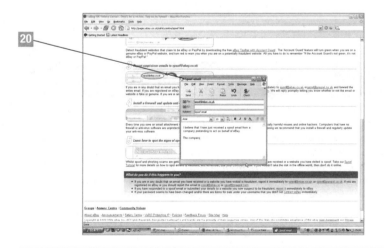

20 And if you do get an example of these damn annoying emails from tricksters, fight back by telling the eBay authorities all about it. Just to repeat, the email address you need to send your missive to is spoof@ebay.co.uk.

Again, a trawl on Google can unearth some useful tools for eBay buyers. You don't even have to stray as far as Google, of course – we've seen eBay's own Buying Resources, and you can also go to http://ebay.volantis.net/anywhere to pick up a useful mobile phone service, which lets you keep up to date with eBay via text messages sent to you about bids, the end of auctions, and whether you've won or lost. The service is called eBay Anywhere and is worth the trouble of investigation if you reckon that your eBay use is going to be pretty heavy.

Whilst not strictly an add-on, don't forget eBay Pulse at http://pulse.ebay.co.uk, which helps you surf the eBay Zeitgeist and find out just what the month's most popular searches and shops are. Meanwhile, if you're after some general tools that are likely to help you whether you're a buyer or a seller, we recommend the list of items at www.auctionlotwatch.co.uk/tools.html. Auction Lotwatch is an excellent site for all auction fanatics.

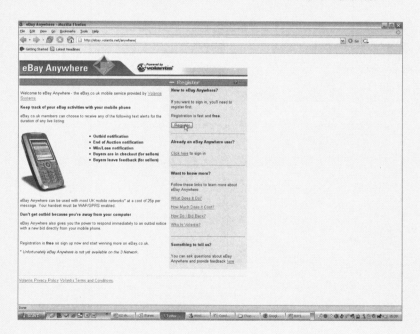

Tools for eBay buyers

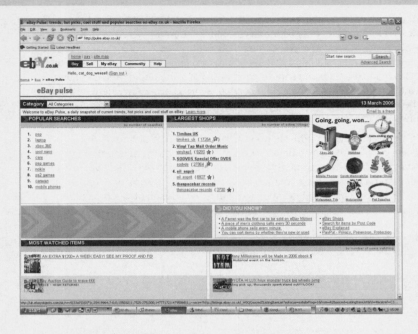

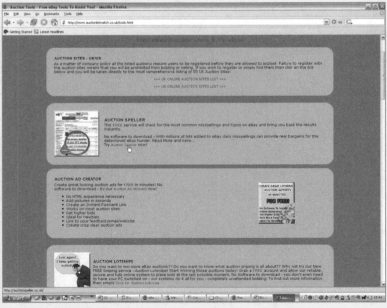

→ Top security tips

eBay's Safety Centre (http://pages.ebay.co.uk/safetycentre/) can help you steer clear of many possible security fears, and warn you about the dangers of spoof and hoax emails. Remember, eBay and PayPal will NEVER send you an email asking you for your account or credit card details, or password. If you get correspondence like this, report it to eBay immediately at spoof@ebay.co.uk. Make sure your computer has got up to date virus protection, obviously – sites such as www.symantec.com are ideal for purchasing new security programs, or using free trials. When it comes to specific auction safety issues, two sites worth taking note of are www.auctionblacklist.com, where you can access a database of fraudulent auctions, and www.swift-find.com, where you can check in seconds whether an item up for sale is legitimate. Don't be paranoid when it comes to online security – just be careful and keep your eyes open.

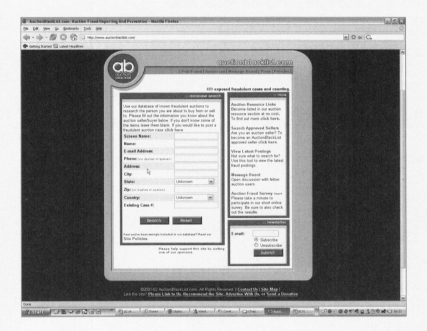

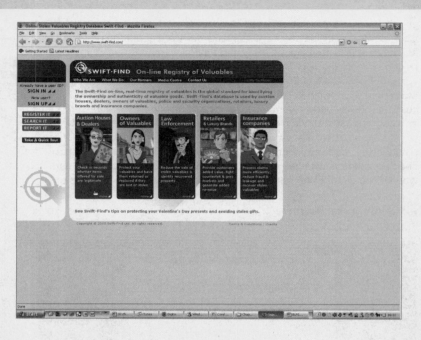

3 Selling on eBay

WHAT YOU'LL DO

→ Open a seller's account on eBay

→ Choose your item to sell

→ Discover different kinds of auction items

→ Create your auction listing

→ Discover different tools to help your listings

→ Understand what makes a good listing

→ Deal with payment

→ Deal with shipping

→ Package your item correctly

→ Learn how to leave feedback

→ Write good feedback

→ Cope when things go wrong

→ Learn about fair feedback

→ Use eBay Selling Resources

→ Use 3rd-party selling tools

What kind of eBayer do you think you are? Are you a casual user, who'll pop in from time to time to browse and try and pick up a bargain or two, or do you think you're more akin to an enthusiastic auction fanatic, who checks back to the site virtually every day, gradually incorporating it into their everyday life? Once you've spent a few weeks on eBay, you'll have a much better idea about the answer to this question – and you should have decided whether you want to be just a buyer, just a seller, or a combination of both.

Most people opt to test the waters, browsing and maybe buying a couple of items, before jumping into selling. If you decide to ignore selling completely, you're missing out on a heck of a lot of fun and, more importantly, countless chances to make seriously significant amounts of cash. Thousands of people, indeed, start off cautiously selling their wares on eBay, gaining experience and knowledge of the whole process, building up their feedback rating and gaining the trust of buyers. All of a sudden they realise that the money is flooding in, and they're getting close to being able to claim the treasured title of eBay 'PowerSeller'.

Let's not get ahead of ourselves, however. The path to becoming a successful eBay seller involves time, hard work, commitment, strong market knowledge and patience when things go wrong. Don't try and run before you can walk – know your limitations, know what you want to sell, and build accordingly. The good news from the very beginning is the fact that because the eBay marketplace is so vast, virtually anything can be sold. One man's tat is another person's gold; you really will be surprised how something you've had festering away in the deepest confines of your loft can spark a bidding war if it's listed effectively on eBay.

You could write a whole book just about selling on eBay – the strategies and tactics you can use are immense and can get quite complex. We're going to focus on the essentials, however – registering an account in the first place, getting a listing online, dealing with issues such as payment, shipping and conflict resolution, and using eBay Selling Resources to get the most from your listings. It's best to see eBay selling as a never-ending road of learning and discovery – and hopefully stacks of cash, of course! Let's get going and begin to explore one of the most popular activities online – becoming an eBay seller.

The first steps in your seller's life on eBay are, of course, to do with opening a seller's account. The good news is that this is incredibly easy; eBay has worked over the years on cutting out the restrictions to opening an account, to the extent that all you really need to be able to do is provide information that confirms you are who you say you are, by submitting details of a credit or debit card belonging to you. No money whatsoever will be taken by eBay until you have an item fully listed and live on its auctions. Let's whizz through the steps you need to go through to be a fully registered eBay seller.

Let's sell!

1. Hopefully by now you'll have registered with the eBay site itself, and had a good look around. If you've made up your mind that you want to be a seller, click the Sell tab on the home page.

Timesaver tip

Before you go to the time and trouble of becoming an eBay seller, take some time out and work out the kind of things that you want to sell. That way, you can get going with the serious business of auction listing as quickly as possible. We'll look at the kind of items that are on eBay shortly, but the quick answer is – pretty much anything you can think of!

Timesaver tip

One of the first things to sort out as you take the initial steps to become an eBay seller is to make sure you have a decent digital camera to hand – either your own or a friend's. It doesn't have to be a state-of-the-art model costing hundreds of pounds; high street stores such as Jessops (www.jessops.co.uk) now offer perfectly serviceable models at wallet-friendly prices. If you can't take a photograph or two of your item that you're putting on online, you really are going to make your life as a seller massively harder. People just won't want to trust you if they can't see what they're buying.

How to pay?

2 Read the Why Sell on eBay? and How Do You Sell on eBay? introductory paragraphs if you like, before clicking Sell Your Item.

3 All prospective members of eBay Selling have to confirm their identity and select a payment method for the fees that the site charges to list items. No money whatsoever will be taken from your account at this stage. Decide whether you want to use Visa or Mastercard, or Switch, Solo and direct debit. The former option is the quickest path to take.

Timesaver tip

Don't underestimate the importance of choosing a good username, on any auction website. Usernames that refer to the kind of items you're going to be selling on eBay, the town that you come from, or your own personality, are handy in being a good reference point for people who want to know a bit more about you and whether they should trust your selling credentials.

Credit card information

4 Now enter your credit-card details, including the number and the address as it appears on your monthly statement. One of the indications that this is a safe site is the 'https://' in the address bar at the top of your browser.

5 Confirm your choice of payment for selling under How Do You Want to Pay Your eBay Selling Fees?

6 Click Continue when done.

Important

If you want some added peace of mind about how eBay uses your personal information, and confirmation of the fact that it doesn't give your details to third parties without your consent, take a look at the Privacy Policy at http://pages.ebay.co.uk/help/policies/privacy-policy.html.

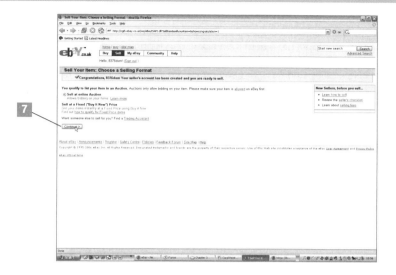

Confirmation

7 eBay will check that your card details match up with who you say you are, and that hopefully should be it! You have now created a seller's account, and can get going with the serious business of researching and then listing your item.

For your information

New to the world of selling, and a bit apprehensive about it? Then pay a visit to the friendly people at the New to Selling help board, at http://forums.ebay.co.uk/forum.jspa?forumID=3002.

Now we enter a slightly abstract, but nevertheless very important phase of the selling process – market research. To put it simply, what are you going to sell? How many other people have sold it on eBay, and for how much? How much can you expect to pay for your item in the high street? Is the condition that your item is in – shiny and new, or used and worn – going to make a difference to how much you can list it for? Is it a rare item which you can expect dozens, if not hundreds, of people to be interested in? All these questions and more need to be answered. If you just go ahead and randomly put 'any old thing' on eBay at any old price, unless you strike it lucky you're going to be wasting precious time, and minimising your chances of making a sale.

The time you spend on this part of a sale is up to you, and, to some extent, depends on how much time in a week you're able to devote to eBay, and also how many items you're putting up for sale – if you're planning loads of sales, researching everything may be a time commitment that you simply can't make. However, when we say that even a few minutes spent on this process can mean the difference between pennies and pounds, you might begin to think that squeezing in a bit of background research may just be one of the most sensible things you ever do on eBay.

Get the pulse

1 eBay Pulse is a popular area of eBay, where you can find out what's hot right now, what the most watched items are and the most popular searches. Regular use of Pulse (which you can get to directly at http://pulse.ebay.co.uk/, or by clicking the link on the home page, as shown here) can give you a vital glimpse into the mind of the eBay population at large, and maybe even give you a clue or two about what it is you should sell.

Research time

2 eBay Pulse can help you see what kind of general items are selling like hot cakes online, but what if you have a specific item you'd like to sell, and want to know what the competition is like? A simple search can help. We're keen on selling a digital radio we got for Christmas, so we type the product name into the Search box under Welcome to eBay on the home page.

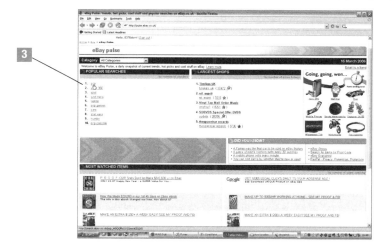

3 Here, the Pulse shows us that the most popular search at the moment is for 'psp' – Sony's sexy new handheld which has been a big hit with gamers over the last year or so. Click on the link to find the very latest listings for that item. We found 20,166 items for 'psp', which goes some way to show the popularity of the item. A popular item can work both ways, of course – it shows that loads of people are interested in it, but on the other hand, do you really want to try and compete with hundreds of other people, selling exactly the same thing? That can be a tad daunting when you're a new seller.

Did you know? **?**

Dotted around eBay are odd snippets to do with sales on the site. For example, a mobile phone sells every minute on eBay, an MP3 player every four minutes, and a teddy bear every two minutes. That's a lot of sales!

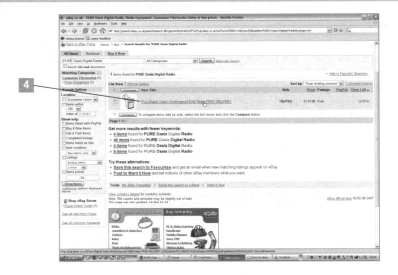

4   There's one result for our exact search, the photo showing us that it is indeed the very same item we want to sell. A few more results below that are close to our search term show to us that there may be some other listings of the same item under different headings, but the results are pretty minimal. We choose to take this as good news, in the sense that digital radio is big news at the moment, so there's a high chance that people will be looking for a smart new model, and there's not much eBay competition around. Click on the item title to be taken to more details about the auction.

Did you know?

eBay Motors is one of the most popular areas of the site, and is an absolute dream for anyone looking for a new vehicle, or just extra parts. By no means is it a resting place for tat, either – the very first item to be sold on the site was a Ferrari!

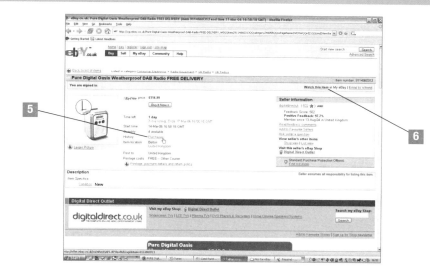

On the hunt

5 The item we want to sell is listed at a Buy It Now price of £114.99 – more than we were planning to list it for. But then our item is second-hand, and it's on sale 'New' here. Click Purchases to see if it's been sold in the past (there are four of the items available in this auction currently).

6 Keep your eye on the auction's progress, and whether it sells or not at the Buy It Now price, by choosing to watch it in My eBay. Click Watch This Item in the top right-hand corner.

Did you know?

Every so often a celebrity auction will appear on eBay and hit the headlines. Auctions in the past have included Lady Thatcher's handbag (which went for £103,000!), a date with GMTV presenter Penny Smith (£9,000), Jamie Oliver's scooter (£7,600) and Joanna Lumley's Ferrari (£35,000).

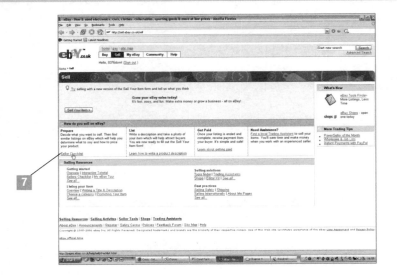

7

Use the Checklist

7 As we've said, how you research on eBay depends on your time constraints and what you're selling. If you're a beginner to selling, at http://sell.ebay/co.uk/sell is a very handy list to eBay's Seller Checklist, which highlights the key things you need to be thinking about when you're listing an item.

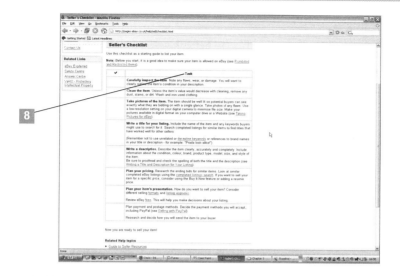

8 Read the list carefully to check that you can comply with all of the recommendations. If you know what you're going to sell, have an idea about pricing, and have this checklist close to hand, you're well on your way to being able to set the wheels of a listing in motion.

→ Different kinds of auction items

When it comes to what's on sale on eBay, where can you start? It's mind boggling. You'll already have noticed the number of categories online, each with several sub-categories as well; what's more, this is just the tip of the iceberg. Click on Featured Items on the home page to get an idea of the sheer scale of the site – we did and got listings for everything from a Mini Cooper S Convertible, to an extended dining set, to an electric guitar, to an 'Allure Leather and Buckle Bustier'. And that's just one page! And then there're the charity listings, 'special and unique auctions', celebrity auctions, eBay Shops, eBay Motors... and the numerous separate eBay sites based all around the world, from Argentina to Taiwan. It is a cliché, but someone, somewhere, probably wants your item, no matter what it is. Knowing that, the next step is to maximise its chances of being sold by setting up an excellent listing, which has a clear title and format, an excellent description, clear photos and a tidy look. And we'll be doing just that next.

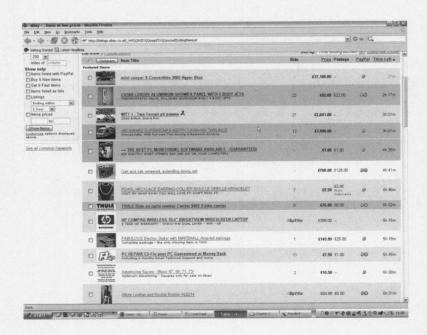

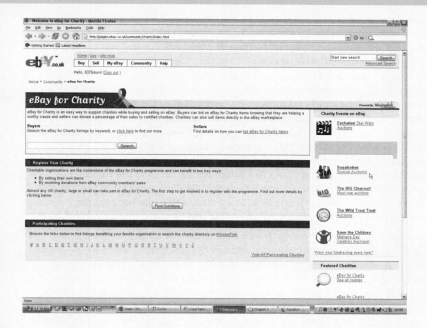

We've now reached the business end of selling – creating your listing. Hopefully, you're already starting off with a clear idea of what you want to sell, a good understanding of how the market works and full knowledge that there's no 'miracle solution' when it comes to selling your wares. You're going to need time, patience, good humour and a willingness to take the rough with the smooth. If you've got all of that sorted, then you can proceed with confidence. eBay holds your hand through virtually every step of the selling process, and rare is the time when you're left floundering around not having a clue what to do next. Despite the fact that actually getting through on the phone to a human being who works for eBay is virtually impossible, the site actually works incredibly well through a kind of self-policing system, and plenty of Help links of course. There'll always be a few rotten apples spoiling the basket, but considering the amount of transactions that take place on eBay every day, the number of disputes is staggeringly low. Plenty of people become prolific eBay sellers with only the bare minimum of fuss and hassle.

Anyway, creating your listing does take a while – or at least it should do, if you're doing things right. You can shorten the process by using selling tools such as eBay's very own Turbo Lister, which we'll look at later, but from a beginner's standpoint, it's best to first spend a healthy amount of time on your listings. You can save listings as drafts, preview virtually every step of the process, and get other people to look at what you've done – any help you can get is crucial when it comes to making that first sale. Below, we go through an example listing, pointing out some good techniques and things to avoid.

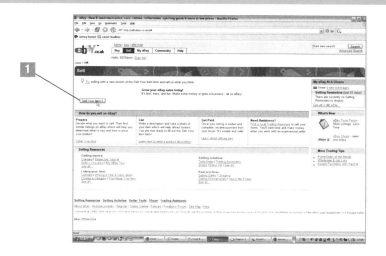

Steps to sell

1 At http://sell.ebay.co.uk/sell, click on Sell Your Item. We're going to sell the digital radio we researched in our last task.

2 As usual, eBay has plenty of help links for you to take a leisurely trip around before you commit to anything, so why not click Learn How to Sell on the right-hand side for some useful pointers?

For your information

Early on in the selling procedure, you'll be asked to decide whether to sell at online auction, or through a fixed Buy It Now price. To be able to qualify to sell through Buy It Now, where you can sell your items instantly at a fixed price, you need to have either a minimum feedback of 10, have confirmed Debit Payment on file for payment of eBay fees, or have a minimum feedback score of 5, a PayPal account and acceptance of PayPal as a payment method. None of these tasks is especially onerous, but if you're a brand new seller and want to get used to the eBay game, you're probably best off plumping for the conventional auction route at first.

Help

3 A help window pops up on the right-hand side, giving you a basic checklist of the steps you currently need to be thinking about in order to get a sale up and running.

4 You may also be worried about selling fees, which will vary from item to item on eBay. These fees help eBay make money out of the whole process of allowing your sale, and are one of the sticks which other auction sites use from time to time to attack the auction giant. Selling fees will affect your profits on eBay, of that there is no doubt, but we're talking small beer really in the great scheme of things, especially if you start having some successful sales. After all, where else can you get a potential audience for your wares of over 10 million people in the UK alone? The Selling Fees link on the right will let you know what to expect when it comes to forking out monies.

For your information

When we were online, there was a new version of the Sell Your Item form, which eBay was obviously trialling to encourage customer feedback and comments. We stuck with the older version for the purposes of this walkthrough, but if the Try Selling With a New Version link is still up when you go to sell your item, feel free to try it and see what the difference is.

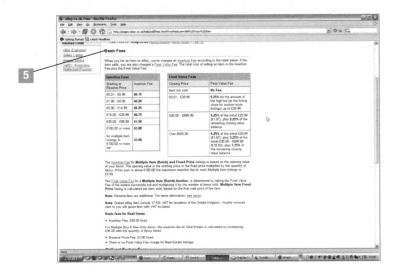

Useful info

5 Basic fees include an Insertion Fee, payable whenever you list an item on eBay, and a Final Value Fee, calculated after your sale. No sale – no final value fee. Insertion fees start from as little as 15p, although you need to take into account the extra costs that can accrue from an auction listing when you decide to go for special features such as bold, highlighting, or making your item a featured listing. Full price rundowns are on this Help page which we accessed in the previous step.

6 Back on the Sell Your Item page, find out more about the online auction process, as opposed to But It Now, by clicking Learn More next to the option.

Important

Is your item actually allowed on the UK version of eBay? Go to http://pages.ebay.co.uk/help/sell/item_allowed.html to check that you're not contravening any listings policies.

Time to list

7 Again, a window pops up giving you the info you need.

8 Right, we've done all the background research needed to be pretty confident about where we're going. Click Continue to enter the first stage of listing.

9 Choose a main category for your item to go into. Fairly obviously, our digital radio goes snugly in Consumer Electronics.

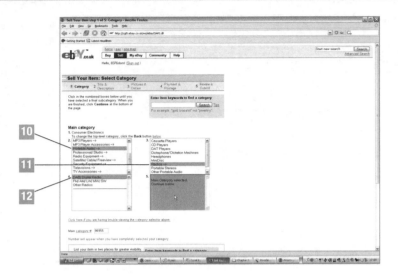

Category breakdown

10 From Consumer Electronics, we can further break down what our item is in the sequential dialog boxes that follow. So Consumer Electronics breaks down into Portable Audio...

11 ...which breaks down into Radios...

12 ...which breaks down into DAB/Digital Radio, which is exactly what our item is. This process is vital in fitting your item in the exact zone it needs to be in, helping it get seen by the right kind of interested parties.

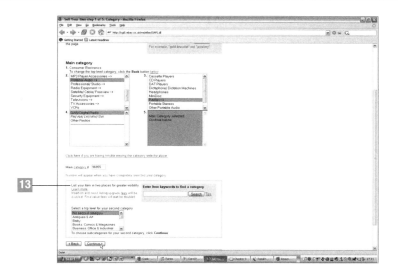

13 If you're cautious about whether you've made the right choice, or keen on getting your item more visibility, you can choose to plump for a second category here. Doing so will increase your fees, though, so be sure that this is something you want to be doing before you go ahead.

Timesaver tip

Choose your category carefully – it won't always be as obvious as our choice was. If you're not sure which category is right for you, enter a keyword to do with your item in the Search box at the top of the Select Category screen. eBay will return with suggestions as to where your item should be, which will help focus your mind and ensure you don't make an embarrassing mistake.

Title please

14 Now choose a title for your item. You have 55 characters to play with. Click Tips on Writing a Good Title if you want help.

15 A window pops up to give you some valuable pointers.

Timesaver tip

Your title needs to be accurate and tell people everything they need to know about the item that's on sale. Look at the difference between 'Oasis Pure Rechargeable Weatherproof DAB Radio', which is what we go for, and 'WOW! Amazing quality sound with this fantastic model!!'. This latter 'description' may at first hand seem to be positive and attention grabbing, but it tells you absolutely nothing about what the item actually is, what it does, and what its major selling point is – which is ultimately the fact that it's a digital radio. We're a bit worried about whether people will understand 'DAB', but those in the know certainly will, and the acronym is becoming more and more widespread in the UK as well. Titles should be clear and concise, honest and descriptive, letting the buyer know exactly what's on offer, whether it's a brand name or not, and what it's major plus point is. If other people have sold your item, see what they've put as their titles as well. Don't underestimate the power of a good title!

Moving on

16 Write a subtitle if you like, to give buyers more info. If you've done the job well enough in the title, however, do you really need this step, considering it costs 35p on top of everything else? We leave it blank.

17 State the condition of your item. Be honest.

18 Now for the all-important item description. Click the Tips Link for Help again if you need to.

19 Read through the Help box on the right about writing an item description.

For your information

Things to include in your item description: the item's features, its overall look and appearance, any blemishes and scratches, what its major advantages and benefits are, how much it cost originally, any relevant background information… anything and everything that allows the buyer to make an informed decision on whether they want to buy what you're offering. Don't waffle, lie, exaggerate or fill your description with tiresome hyperbole and exclamation marks. If you think it's important, put it in, and remember you're not writing an essay. Show the description to a pal to get their opinion on whether you've said everything that needs to be said in as economical and effective a way as possible. The more things you sell, the more you'll grasp just what is needed at this stage of the game.

Get descriptive

20 Here's our description under way. Because it's a technology item, we can get away with a bit more jargon and item specifications, and what makes this DAB radio different from the rest. We start to list the features of the item.

21 Hopefully our list of the salient features will convince a few people that this is one of the best DAB radios on the market.

22 Go for a spell check to avoid any embarrassing typographical bloopers.

Important

If you save your listing as a draft, and then start a new listing, your draft will be deleted, which could mean you lose hours of work. So keep your eye on your listings, and make sure that you do actually complete one sooner or later, rather than leaving it hanging for too long.

Checking

23 eBay will recommend any changes that it sees fit. If you agree with it, click Change to carry out an alteration.

24 Click Preview Description to see how it'll look to buyers. It's a good job we did, as we'd forgotten to enter the <p> command in our description, which tells eBay to insert new paragraphs where necessary, rather than just run everything in an untidy couple of lines.

25 When you're happy, click Continue.

Timesaver tip

Pay attention to spelling in an item description. No one will be impressed with an item that has spelling mistakes everywhere; it reflects badly on the seller and could mean the difference between a possible sale and a rejection.

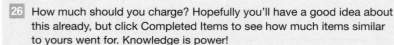

Details

26 How much should you charge? Hopefully you'll have a good idea about this already, but click Completed Items to see how much items similar to yours went for. Knowledge is power!

27 Now we're onto pricing and duration. What's your starting price? Do you want to donate some of your sale to charity? How long is the auction going to last? Do you want to set a reserve price? Where is the item's location? Decide here.

For your information

The reserve price is the lowest price at which you're willing to sell your item. If the auction doesn't reach this price, there'll be no sale. You don't have to set a reserve price, and if you do, there's a fee. And you must set the reserve value to at least £50.

Important

If you want to be really strategic, or set the auction closing time to a time of the day when you know you're going to be around, you can select Schedule Start Time and enter an exact time for things to kick off. This will cost you an extra six pence.

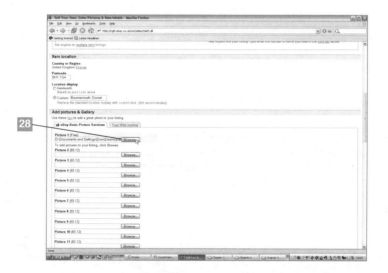

Photo crazy

 Now use eBay Basic Picture Services to add an all-important picture of your item to the mix. We're going to add one picture at the moment, and then work out our prices to see if we should add more. The first picture is free, additional pictures are 12p each.

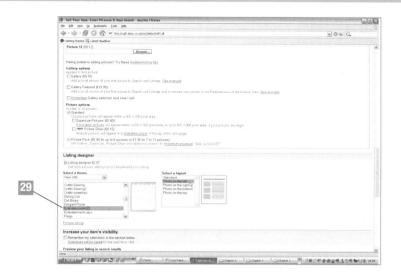

29 You can also choose Gallery at 15p, which adds a small version of your picture for people to see when they search listings. Choosing Gallery Featured, where people can see your pic in the Featured part of the gallery, will set you back £15.95, but really helps sales. Confirm your other picture options (standard or extra large pics?) and then enter the Listing Designer, where you can choose a theme and a layout for your listing, at a cost of an extra 7p. We go for a suitable music-related theme.

For your information

At http://pages.ebay.com/help/sell/photo_tutorial.html, eBay has its own list of useful tips for taking photos. As we've mentioned ourselves, the importance of having a digital camera is stressed – with an amusingly cheeky comment recommending that if you don't have one, you snap one up on eBay! Other recommendations include using proper lighting to take your photos, photographing at close range and at an angle, avoiding over-reliance on flash, and setting the camera to a medium resolution (1024 x 768 pixels).

For your information

Remember to save photos for use on eBay as 'jpg' files. You can choose to supersize your pictures for an extra 60p, or get a slideshow player going on your auction for an extra 15p.

Promotion

30 If you want to make your item more visible, you can pay eBay to carry out a number of font tricks for you, including bold (75p) and highlighting (£2.50). Very useful, but don't forget they add to the total cost of everything. It's unlikely at this stage that you're going to want your item to be in a featured category (£9.95) or featured on the home page (a whopping £49.95! Think of the exposure, however) – both these options require you to have a feedback rating of 10 or more as well. Make your choices and click Continue.

31 Sort out your payment methods on the next page – what will you accept from buyers? And then state where you're willing to post the item to, clicking Learn More About Posting Internationally if you want some background info.

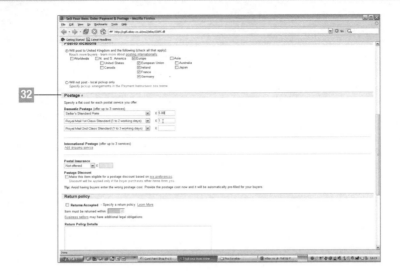

32   Then state your postage costs. If you're not sure what to do here, check eBay's help section for tips, or your local post office. Obviously the bigger the item, the more you're going to have to think very carefully about postage and packing. We'll look at these areas in closer detail later in the chapter.

Timesaver tip

The principle of being as honest as possible about your listing applies to photos as well as text. Don't try and cover up any blemishes, tears and scars on your items – indeed, you may want to take photos of exactly these kinds of issues, just to give your buyer a full picture of what they're going to be getting. Any nasty surprises that the buyer is faced with when they get the item could come back and bite you on the hand when it comes to feedback, and the chances are that if you photograph your item truthfully and accurately, the buyer will be impressed and more convinced that you're someone worth doing business with.

For your information

One of the best programs for an eBayer on a budget, who wants to be able to tweak their photos effectively and get them into tip-top shape for the web, is Photoshop's little brother, Photoshop Elements. You can find out more about the program on plenty of sites on the web, including www.adobe.co.uk/products/photoshopelwin/main.html.

33 Are you going to have a returns policy? Do you have any payment instructions for the buyer to take note of? Sort these details out here, before clicking Continue.

34 Things should be falling into place now on this, the Review & Submit final stage of proceedings. Virtually everything you've done so far can be changed by clicking on the Edit button on the right – it's vital you make time here to go through everything with a fine-tooth comb.

35 Click Preview How Your Item Will Look to Buyers for an invaluable preview of what you've done.

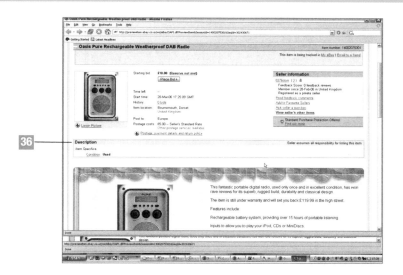

Final steps to submit

36 You can now see how it looks. Not satisfied? Then go back and edit until you are!

37 Once you're 100 per cent happy, and only then, click Submit Listing. Your item will be listed on eBay, the fees you've accrued (which are detailed in a summary table above, will be charged to your account, and it'll be all systems go! As you can see, we've racked up over £6 in listing fees, without even really trying, and certainly without being flash. The buzz you get through putting your item online and waiting for people to bid on it is one of the most rewarding elements of eBay – you'll never forget your first bid, or your first sale.

→ Different tools to help your listings

We'll look at eBay's Selling Resources later in the chapter – they're as useful as the corresponding Buyer tools. Elsewhere, we've mentioned how you can give your listing a leg up by promoting it in the Featured Category or Home Page areas – at a high cost, however. Slightly more modest options such as the 15p Gallery option may be more suitable, or a free Andale counter to track how many visitors your listing has (Andale's home page is www.andale.com). Striking a balance between effective eBay listings features, and still keeping costs down, is an important eBay skill to learn.

3

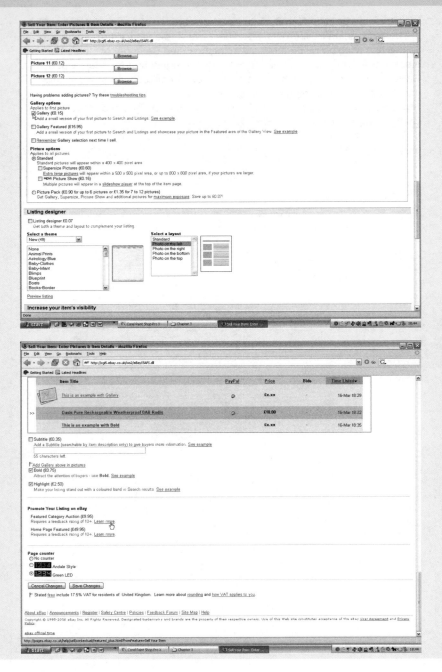

→ Examples of good and bad auction listings

We're not going to name and shame anyone here – everybody makes mistakes and you'll certainly see many on eBay. It is frustrating, however, to see countless get-rich-quick scams: auctions where a picture of a scantily-clad woman is used as bait to attract attention; insulting, rude or profane listings; spelling howlers such as a sale for a 'plastation'; atrocious photography; listings in eye-shatteringly garish colours and fonts, where you're SHOUTED AT with loads of EXCLAMATION MARKS AND CAPITAL LETTERS... the list goes on. For all the bad auctions, though, plenty of excellent, professional listings exist – we were especially impressed in our travels with a listing for a caravan, which had all the photographs and item specifications you could want, and a sale of an Audi A4 1.9L TDi SE 130 BHP which was extremely thorough and well-done. What to us makes a good listing? Lots of photographs, if they're needed, calm, well-written, accurate prose, a desirable item and a seller who you can contact at the click of a button. Everyone's tastes will vary, but click around eBay for a while and you'll soon be able to spot the difference between a dreadful auction and one that really takes off and impresses.

3

Important

Dos and Don'ts of auction listings
To summarise some of our findings, then:

Do:
Research how your item has sold in the past on eBay.
Be as honest and upfront as possible about your item, telling the buyer everything they need to know, spell-checking thoroughly and stating explicitly the condition of what you're selling.
Take good, well-lit, accurate photographs, possibly from more than one angle (especially with items such as cars).
Understand all the costs you're likely to incur, including listing fees and postage and packing.
Be patient and don't jack it all in if your first few listings don't attract bids. Learn from your mistakes and act on them.

Don't:
Lie in your listings, or cover up the condition of an item with vague language or craftily taken photographs.
Litter your listings with spelling errors and grammatical howlers.
Price your item unrealistically.
Litter your listing with bells and whistles that end up costing you an arm and a leg.

If you've gone to all the trouble of setting up an auction, and had the skill and good fortune to engineer a successful sale, you're going to want the issue of payment to be sorted out as quickly as possible.

As you might expect, eBay does its level best to make the process as swift as possible. When you list an item, the eBay Checkout option is by default turned on. eBay Checkout means that if you sell an item, a Pay Now button appears in the buyer's Items I've Won section of My eBay (and on the item listing itself). The buyer can then call up a page which calculates the fee payable to you, the seller, and then choose how to pay. Payment method will be subject to any demands you placed in your listing.

If the buyer pays by PayPal, payment can be swift and efficient, with the money sent securely to your PayPal account. You can then keep the finances in your own PayPal account, or transfer them to a bank account.

That's the simplest process explained in a nutshell; as you'll see over the next few pages, there's no reason why online payment should be something that worries you.

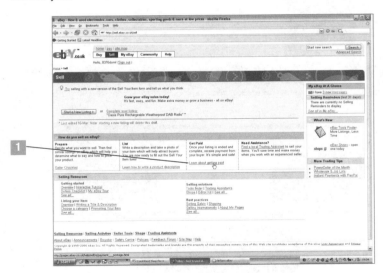

Payment help

1 Back at http://sell.ebay.co.uk/sell, click on Learn About Getting Paid to understand just how everything works.

For your information

You can disable eBay Checkout if you'd rather go down a different path when it comes to accepting monies. When you're using the Sell Your Item form, on the Enter Payment and Postage page and in the Payment Methods section, click Edit Preferences, and turn off the Use Checkout button, which will be enabled by default.

Overview

2 The best thing to go for is a general description of the whole process, so click on Overview from the next page.

See also

Pay as many visits as you like to www.paypal.com/uk to understand just how this immensely popular payment scheme works, and don't forget chapter 5 of this book, where we go through the simple signup procedure.

Timesaver tip

There's no point complaining that the payment process hasn't been as smooth as you would have liked, if your own payment instructions in your auction listing aren't 100 per cent and explicit. Make sure the buyer knows exactly what payment methods you accept, whether you have a preference for one method or another, and whether there are any extra charges that the buyer needs to be aware of. You don't want to be in a position where you're solving loads of problems after the auction has closed, because that way you could be tempting the buyer into leaving you some seriously bad feedback, which will have a knock-on effect for your future sales.

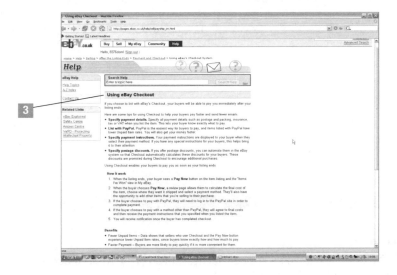

Cool Checkout

3 Gen up on how eBay Checkout works on the next page.

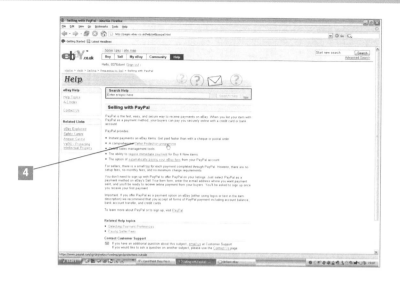

4 Go back a page to where you initially clicked Overview, and now click on PayPal for Sellers, to find out more about this invaluable seller's tool. We opt to find out more about the Seller Protection Programme.

Important

It's no surprise that eBay recommends you list your items with PayPal, as they're partners after all, but the truth is if you're a new seller, or even a pretty experienced one, PayPal is the best method around of accept money. Its security levels are very impressive, and it's one of the speediest methods around for getting the money you're owed. You don't have to sign up with PayPal to list it in your items – just enter a valid email address, and you can deal with signing up once you've achieved your first completed sale.

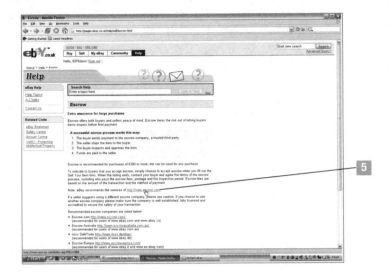

Escrow

5 Go back a page again and click on Escrow. Escrow payment is useful for purchases of more than £250, although you can use it whenever you like. Under Escrow, the buyer sends their payment to a trusted third-party company; the seller then ships the item to the buyer, the buyer inspects and approves the item, and funds are then paid to the seller. eBay recommends a list of Escrow services at http://pages.ebay.co.uk/help/sell/escrow.html, and warns against using untried or untrusted alternatives.

Important

You will be charged a small fee by PayPal for payments completed through the system. Check the website for details of what you can expect – it obviously depends on the final price of the sale. At least PayPal doesn't charge setup fees or monthly fees, which you may experience with other payment providers.

Important

Find out how PayPal offers you protection as a seller at
https://www.paypal.com/cgi-bin/webscr?cmd=p/gen/protections-outside.
Methods include providing free fraud-prevention tools, and acting as a mediator
in a financial dispute between buyer and seller.

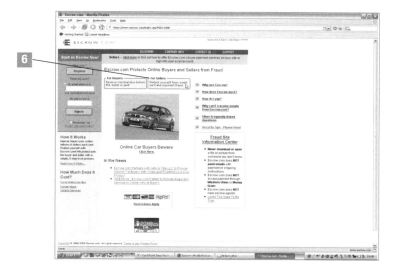

Excellent Escrow

6 We go to www.escrow.com to see what the benefits of the system are.
It's useful to know how the programme works, as it may come in very
handy in your future eBay transactions.

Important

Escrow is not unusual in the fact that it does attract cunning fraudsters who try
to use the system to dupe people. A good webpage to read about Escrow fraud
prevention and how the scams normally work can be found at
http://www.lookstoogoodtobetrue.com/fraud/escrowfraud.html.

7 Complete your crash course in issues surrounding online payment by making full use of the eBay Help area (http://pages.ebay.co.uk/help). Here, our simple query about 'payment' gives us an invaluable rundown on issues surrounding the successful acceptance of online payment.

Important

Dos and Don'ts of accepting payment

A sensible, clear-headed attitude when it comes to online payment is important. If something looks dodgy, then it probably is. Is the buyer trying to push you into accepting payment via an Escrow service you've never heard of? Is the postage information they've provided the same as what's listed in their contact information? If not, why not? Does their feedback rating include any negative comments about their ability to pay? If you have doubts, for whatever reason, about the validity of someone's payment method, or their ability to pay in the manner that they say they will, contact eBay and let them investigate for you. If things go as far as you being unfortunate enough to be on the wrong end of a fraudulent payment, or no payment at all, look at the options of contacting the police and the payment issuer to clear things up. The Contact eBay links in the Help area of the site are numerous, so don't just think that the auction giant will do nothing. You will need to be proactive, however, and explain what's happened clearly and concisely. We'll be looking at dispute management in greater detail later in the chapter, but please don't be put off selling on eBay by us looking at these 'worst case' scenarios. Being aware of what you can do when things go wrong can be a real boon, helping you not to panic, and to avoid such difficulties in the first place through calm, conscientious and through selling methods.

Under the banner of 'best practices' comes how to get your item to the buyer in the best possible condition. This aspect of eBay life is vital if you want to get good feedback and encourage people to deal with you. Plonking your item randomly in a cheap padded envelope or using cheap, inferior and poor-quality parcels will infuriate your buyer, who may demand a refund, or at the very least leave you poor feedback.

Luckily, the web can help you in the task of getting packaging right. The Royal Mail's site (www.royalmail.com) explains some options; packaging suppliers such as Staples (www.staples.co.uk) and Viking Direct (www.viking-direct.co.uk) can help too. Don't cut corners on packaging – it may be worth raising the prices of your auctions, if it means that you can provide a superior parcel. Spend a lot of time thinking about the kind of item you want to post, and how it can best be wrapped up. This kind of attention to detail can prove a real winner in the long run, and we'll list some of our best tips over the next few pages.

Getting shipshape

1 Back at http://sell.ebay.co.uk/sell, under Best Practices is a Shipping link which will get you off on the right foot in this vital area of eBay selling.

2 The page at http://pages.ebay.co.uk/shipping has some top-notch shipping advice, including a host of useful links for you to explore. We click on the Royal Mail logo.

Important

If you're selling more than one item, and they don't all fit in the same parcel, then you're going to have to bite the bullet and send more than one package. The danger of parcels splitting in the post and collapsing is too great for you to take any chances here.

Tips and tricks

3 The Royal Mail site is packed to the rafters with useful information, which you can explore at your leisure. If you're anticipating sending quite a large volume of post over the next few months, for example, under Working Smarter click High Volumes of Mail.

Important

You may find that following best practice packaging tips makes you re-evaluate how much you list your items for. This is fine, however. Buyers will far rather spend a little bit more on an auction where the items are going to reach them in immaculate condition, than get a 'bargain', only to receive something that looks like it's been for ten rounds with Mike Tyson on its way to them. Remember to think ahead, and realise that the buyer can get their revenge with a few well-chosen comments in a feedback review, which could haunt your eBay experience for ever.

How to save

4 Then find out how best to save money on packets and parcels under the Savings link. The Royal Mail is well aware how many people use its services to send eBay items, and many of its services seem to be designed specifically to help this sudden new group of fanatical mailers.

For your information

If you're sending very precious or fragile items in the post, consider whether recorded or registered delivery needs to be employed. Again, this may have an effect on your costing schemes, but it's worth it for the peace of mind that it brings.

Set the Standard

5 The Standard Parcels link may be more to your taste; the site describes this service as easy-to-use and low-cost. It also comes with a free certificate of posting from your local Post Office, which means that it is 'very popular with users of eBay and other online auctions'. Explore how it works so you can see if it's the service you want to use.

Important

Remember to protect your item on all sides when you post it, with a good system of bubble wrap or polystyrene. If you combine this with a strong, sturdy outer box, you're well on the way to protecting yourself against the worst that the postal system can throw at it.

→ Essential packaging tips

To summarise what we've learned about packaging: Remember to match the packaging you use to the item you're sending. Think about using a sturdy outer box that can be closed securely, with the address and return address information written clearly. Make sure you seal the box carefully, with strong tape. Don't overload your package. Use bubble wrap if it will help your package stay secure and not move around. Check out the websites of companies such as The Royal Mail and Staples, as well as eBay's own packing and shipping Answer Centre board (http://answercentre.ebay.co.uk/forum.jsp?forum=8013), and third-party help sites such as www.easyprofits.co.uk/ebay-postage.php. Get this 'dull' aspect of eBay life right, and you'll receive acres of positive feedback, helping you grow and grow in your life as an eBay seller.

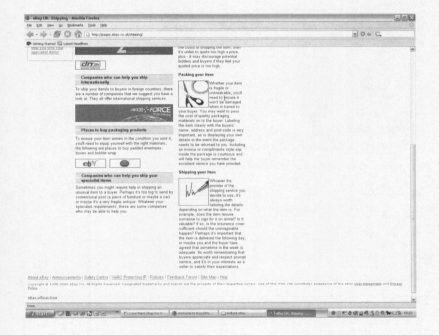

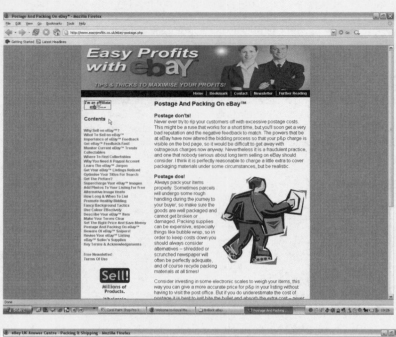

Hand in hand with successful selling, good shipping and a friendly and efficient buyer comes the issue of feedback. If your buyer is fast, efficient and easy to communicate with, leave them feedback and say so. They should return the favour if your auction, payment handling and shipping procedure has been up to scratch.

1 An easy way of leaving feedback is to go to My eBay and click on the Feedback link on the left of the page, then Leave Feedback. From there, you can see who you have had dealings with over the last 90 days, and leave them feedback. Again, the Help section has plenty of tips for getting a proper understanding of feedback.

2 In My eBay, click 'Feedback'.

3 You can see our complimentary comments on the right-hand side of the screen.

Important

Ropey old sticky tape from the back of your kitchen drawer just won't do when it comes to sealing up your item. Get searching online, or in high street stationery stores, for strong parcel tape that will withstand all the things that our great British postal system can throw at it.

Leave it

4 Simply click Leave Feedback to then leave your comments on a recent transaction. Remember, a prompt and fair buyer deserves positive feedback just as much as an efficient and professional seller.

→ Writing good feedback

The basic principle here is to remember just how important feedback is, and
what a crucial role it plays in the successful running of eBay. If something bugs
you during a transaction, contact the buyer (or seller) first by email or phone to
try and clear it up – most people will be very willing to do so. Don't go in a rage
to your PC and type something offensive, derogatory, factually incorrect or
insulting. Anything libellous or defamatory could even land you in hot water
with the law – it's wise to think of this worst-case scenario if it makes you
understand just how important feedback is. When you write feedback, keep
your comments fair, factual and honest, firmly to the point and not waffly.
Feedback you leave is permanent, unless it matches the criteria for abuse
removal, and can be seen by the whole of the eBay community. If you want to
reply to feedback received, use the Feedback Forum in the Community section
of eBay at http://hub.ebay.co.uk/community.

Writing good feedback

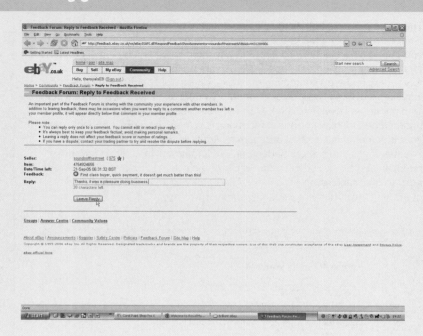

→ Cope when things go wrong

There's no point in denying it – things can go wrong on eBay. Just like in real life, there may be some people out there ready to pull a fast one on you, try and con you out of some cash, or generally put a spanner in the works and prevent what we all want from the site – a fast, easy transaction with everything wrapped up as quickly as possible. Ultimately, however, if botched-up transactions kept happening on eBay, with thousands of rogue buyers or sellers, eBay wouldn't have remained as enduringly popular as it has. If eBay's Dispute Management procedures weren't up to scratch, an increasingly computer-literate British public simply wouldn't put up with it, and would move elsewhere. So the reality of eBay life is clear – there's always the possibility that things might go wrong, but if you hold firm, don't panic, and allow eBay's own services, checks and balances to ride into action if things look like getting serious, the overwhelming likelihood is that things will work out OK. We're not going to make rash promises and tell you that everything will always get sorted out to your satisfaction – life doesn't work like that, and many people do have the odd eBay tale of woe or two. However, if you follow some of the action points over the next few pages, and hold your nerve, you will certainly significantly reduce the chances of being caught in an eBay nightmare.

3

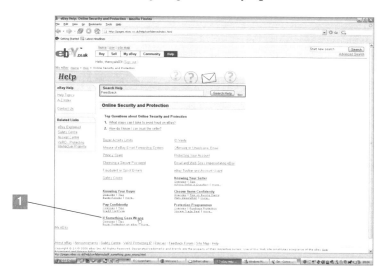

No need to panic

1 From the Help index on eBay at http://pages.ebay.co.uk/help/index.html, try clicking on Online Security and Protection, followed by If Something Goes Wrong. This is a good starting point to understand some of the solutions that eBay has put in place to try and cope with transaction problems.

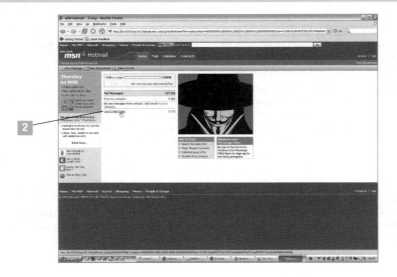

2 The initial overview goes into some of the general things you can do both as a buyer or a seller to try and minimalise problems. Communication is the key, of course: if you have questions during a sale or in the post-purchase procedure, the two parties involved in the deal really need to talk to each other and iron out any uncertainties. People will always have questions, no matter how clear your policies and listing descriptions are, so check your email account frequently, and don't forget to watch those email spam filters carefully. You could be sitting there fuming because the person you're trading with hasn't got in touch with you, only to find that their email is somehow nestling in the Junk folder of your email program. Don't jump to conclusions without knowing your facts.

Important

What, ultimately, can eBay do in disputes? It can investigate cases and close members' accounts if they are found to have abused site policies; it works with PayPal to help with transaction problems, it offers purchase protection (see https://www.paypal.com/uk/cgi-bin/webscr?cmd=_pbp-info-outside) and co-operates with law enforcement agencies in worst-case scenarios. eBay will not contact a member directly to enquire about an item, and also claims (at http://pages.ebay.co.uk/help/confidence/isgw-fraud-ebays-role.html) that 'since we are not involved in the actual transaction, we cannot force a member to live up to their obligation'.

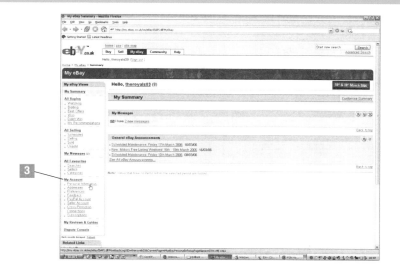

Protection

3 On the theme of encouraging healthy, swift communication, make sure
your contact details are totally correct by checking your details in the
My eBay section of the site, under My Account. Anything that's wrong
in this vital area of your eBay life could cause untold problems when it
comes to dealing with people efficiently.

4 Don't forget eBay's partnership with PayPal. If you sign up with PayPal and go to My Account, you can choose to use the Resolution Centre to file a claim that you've been treated unfairly as a buyer. And for sellers, which is what we're concerned about in this chapter, take a look at https://www.paypal.com/uk/cgi-bin/webscr?cmd=xpt/ua/PolicySpp to see PayPal's Seller Protection Policy, which ultimately protects you against claims by buyers, of unauthorised payments, or 'against claims of non-receipt of any merchandise'.

Important

Again, we must stress that at first it's best not to jump into heavy-handed dispute claims, or start bandying about accusations that have little or no basis in fact. So many problems can be solved with a calm head. At the very beginning, check your buyer's feedback – if they have a long history of happy eBayers giving them glowing references, then that's another indication that it will be best to hold your horses before jumping headlong into an argument.

Unpaid items

5 As a seller, one of your worst nightmares could be getting involved in an unpaid-item process, details of which can be found at http://pages.ebay.co.uk/help/tp/unpaid-item-process.html. Communicate with your buyer as much as possible before entering a process where you file an unpaid-item dispute. Normally you have to wait until seven days after the listing has closed to file this claim through eBay – you can see the link here to start proceedings off. eBay will then contact the buyer with a reminder to pay.

For your information

If the buyer still doesn't pay, and there's been no mutual agreement between the two sides not to complete the transaction, as a seller you can file for a Final Value Fee Credit, which is basically a refund. If you follow the steps carefully, you should get a refund within 48 hours, and may well then be able to relist the item for free.

File it

6   On requesting to file an unpaid-item dispute, you'll be taken to this screen. Check that you satisfy the conditions mentioned on the page, fill in the item number of your listing, and then click Continue to proceed.

For your information

eBay also has a dispute management partner, called SquareTrade, who can help sort out difficult problems and disputes. See what they have to offer at www.squaretrade.com.

Important

If you feel that your eBay account has been illegally accessed by someone, find out what steps to take to try and resolve the situation at http://pages.ebay.co.uk/help/tp/isgw-account-theft-reporting.html.

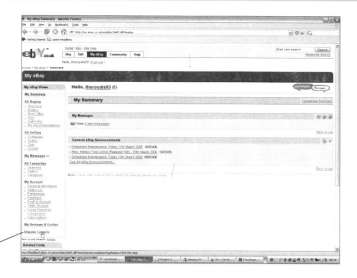

7

Dispute Console

7 The invaluable My eBay can once again be your friend here. In this section of eBay, you can click on Dispute Console to see a list of disputes you have open, along with links to report an unpaid item (as a seller), or report an item that has not been received, or is significantly different from what was described (as a buyer).

For your information

PayPal's Customer Support area can be found at https://www.paypal.com/uk/cgi-bin/webscr?cmd=_contact_us. If a question you have can't be answered here, try phoning them on 08707 307 191 (Mon-Fri 6am-10pm, Sat 8am-10pm, Sun 9am-10pm).

Find them

8 And don't forget eBay's very useful 'Find a Member' page at http://search.ebay.co.uk/ws/search/AdvSearch?sofindtype=8, where you can check an elusive member's status by entering their User ID or email address. One step up from that is getting contact information from someone that you are doing business with (you must be involved in a current or recent transaction with them to request this), which you can do at http://search.ebay.co.uk/ws/search/AdvSearch?sofindtype=9.

→ What is fair feedback?

All these problems we've hypothetically discussed may leave you thinking that there must be an awful lot of negative feedback flying around eBay. Take a look at http://pages.ebay.co.uk/services/forum/feedback.html to understand that feedback should not be entered into lightly, and that anything which is libellous or defamatory can be dealt with relatively swiftly. We've already discussed how to reply to feedback; if you leave negative feedback and then regret it once a deal has been completed, why not try and set in motion a mutual feedback withdrawal? If both the buyer and the seller agree that a feedback comment is not longer fair, initiate this process at http://feedback.ebay.co.uk/ws/eBayISAPI.dll?MFWRequest. It will have to be a pretty exceptional case for eBay to intervene itself to remove any feedback – see http://pages.ebay.co.uk/ help/policies/feedback-abuse-withdrawal.html for some pointers as to when it will.

One of the best things about eBay is that you don't have to stray away from the site to get everything done. When you take into account the brilliant Help section, the friendly Community and forum and the wealth of Buyer and Seller's tools, you've got pretty much the whole package. Even when you need to think about issues such as payment, links from eBay guide you into the arms of PayPal with the minimum of fuss and bother. In the previous chapter, we looked at the Buying Resources that eBay provides you with, including the Reviews and Guides section, the Safety Centre, and the eBay Toolbar. Here, we're going to see how eBay helps out sellers, which pretty much boils down to providing you with the Turbo Lister and Selling Manager tools. Turbo Lister is a free-to-download bulk-listing tool which helps you sell multiple items quickly and efficiently; it can be a real boon if you're trying to get as many professional-looking listings up and running on eBay in as short a time as possible. After all, we all lead busy lives, and anything that can take some of the time out of creating numerous eBay listings should be welcomed with open arms.

Selling Manager, meanwhile, works with My eBay to help you 'sell smarter', with email templates to send out to buyers, facilities to let you reschedule pending listings, and professional-looking labels and invoices to print out.

Let's dive into the Selling Resources over the next few pages.

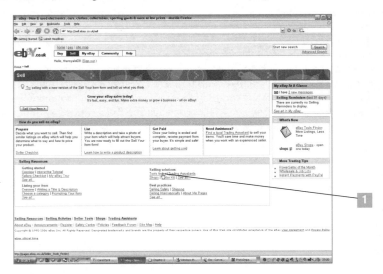

What tool is right for me?

1. eBay has a few selling tools on offer, so to find out which ones are going to be most useful to you in your current position as a seller, click on Tools Finder under Selling Solutions at http://sell.ebay.co.uk/sell once more.

2 In the run-up to the ultimate aim of becoming a privileged eBay PowerSeller, effective use of selling tools can make all the difference. To see which tools are right for you, answer the three questions in the boxes here, with your answers leading to each new set of options. We are currently selling unwanted items occasionally, want to improve our listings look and feel, and would like a variety of templates to help us achieve our aims.

3 The Sell Your Item Form is recommended, not surprisingly, but also Turbo Lister. We highlight it and then click See My Tool Choice.

For your information

eBay also has a list of 'Trading Assistants', who can help you save time and money by actually selling your item for you. Obviously they'll take a cut of the profits for their trouble. Find one near you, who possibly trades in the kind of thing that you're likely to be selling, at http://tradingassistant.ebay.co.uk/ws/eBayISAPI.dll?TradingAssistant&page=main.

For your information

In a similar vein to Trading Assistants is the rise of 'Drop-off Shops' in the UK. If you haven't got time to do your own eBay listings, deliver your goods to these shops, wait for the experts to give them the green light for selling, and then watch as they put up a professional listing of your item, and then deal with all the business of bids and postage to the successful buyer. Obviously you'll still be charged listing fees and a commission fee, but this is a great method to sell stuff without doing much work at all. The main players in this market include SellStuffEasy (www.sellstuffeasy.com), Auction Assist (www.auctionassist.co.uk) and AuctionSeller (www.auctionseller.co.uk).

Turbo Lister

4 A wealth of info about TurboLister appears, telling you how this simple tool can help you get a host of listings in tip-top shape in a matter of minutes. It's a desktop tool, downloadable in seconds, which can really enhance your selling life. We choose to take a tour of the program. Click Begin Preview to start.

5 Cycle through the tour to get a good understanding of just what Turbo Lister can do for you. When you've finished, click Home.

Important

Turbo Lister does not currently support Macs.

Timesaver tip

If you're going to be juggling lots of separate auctions, TurboLister is ideal, as it lets you schedule your listings to start at a specific time. You can then work out a programme of staggered listings, so you're not overwhelmed by dozens of transactions all at the same time.

Let's download it

6 Happy that Turbo Lister can help you out in your selling life? You might as well download it – it's free, sits unobtrusively on your desktop, and you can always press it into action after a few weeks when you're more confident about your listings and what you want to sell. Back on the Turbo Lister home page (http://pages.ebay.co.uk/turbo_lister), click Download Now!

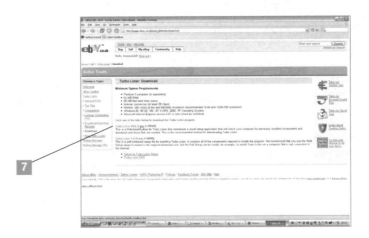

7 Check that your computer is set up for Turbo Lister, then choose a download link.

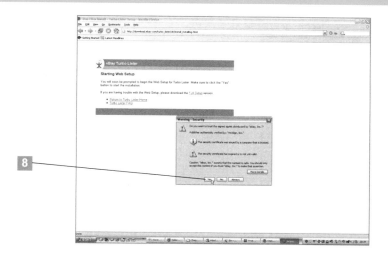

8 Click Yes when prompted to begin installation.

Downloading

9 TurboLister is now preparing to install. Depending on the speed of your internet connection, you should only have to wait a couple of minutes at most for this process to complete.

10 Once installation has finished, tick the check box next to Yes, I Want to Launch Turbo Lister Now, and click Finish.

For your information

Your computer should have no problems running the Turbo Lister application. You'll need 20 Mb free hard-drive space, 64 Mb RAM, an internet connection (obviously), Internet Explorer 5.01 or later, and an Operating System running one from this list: Windows 98, 98 SE, ME, NT 4 + SP6, 2000, XP.

For your information

Click on Help, Help Topics in Turbo Lister to get useful advice that will make you understand how the program works.

For your information

From time to time there may be updates released of Turbo Lister. To
check for these, go to Check for Updates Now in the Advanced
Options section of Turbo Lister.

3

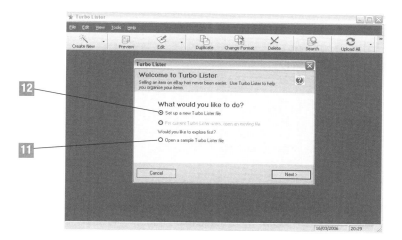

First steps

11 The program loads, with a blank interface. You can choose to explore
the program by opening a sample Turbo Lister file, to see what a listing
looks like.

12 Or set up a new file if you're confident to learn a little bit on the hoof.

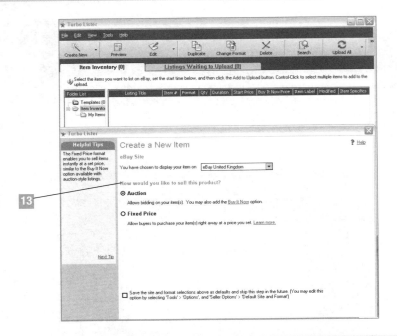

13 The listing screen is in the top half of this screenshot; below is the Create a New Item setup where you can go through the steps to get your listing up and running in Turbo Lister. See our tips dotted around these pages for more specific information about how Turbo Lister ticks.

For your information

To create a new listing in TurboLister, click Create a New Item, choose your auction format, enter a title, choose your category, and set about designing your listing. Just as with the Sell Your Item form, you can see a preview of every step you take in TurboLister. Then set about adding the pricing and postage details for your listing, before clicking Save when you're totally happy.

For your information

When you have several listings up in TurboLister, simply click Edit in
the toolbar on a highlighted item to change its details. Managing the
start times of your pending listings is a crucial skill to learn; in the Item
Inventory View of Turbo Lister, simply select a pending listing and click
Reschedule. Then enter your desired new start time.

Selling Manager

14 Maybe Turbo Lister isn't enough for you – what else can eBay offer?
Well, back at the Selling Tools Finder screen (http://pages.ebay.co.uk/
Seller_Tools_Finder), try clicking on Selling Manager or Selling Manager
Pro on the left-hand side, to see what these two eBay-backed utilities
can do for you. We opt for Selling Manager, and a tour again to see
what's what.

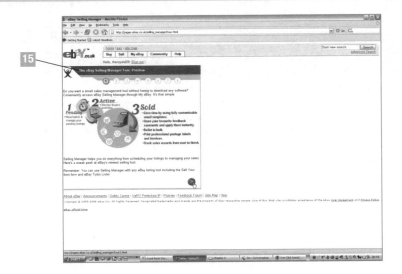

15 Selling Manager doesn't even require you to download any software –
it's all accessible via the ever-useful My eBay section of the site. Agree
to use this utility, and it tacks itself on to your existing My eBay set up,
adding a whole new Selling Manager tab to your list on the left-hand
side. Selling Manager's role in life, to let you see an easily digestible
summary of all your scheduled, active, unsold and sold items, along
with what's awaiting payment, awaiting postage or paid and
dispatched, as well as provide you with the ability to print professional
invoices and postage labels, should not be underestimated. If you're
serious about selling, you need Selling Manager, and the basic version
of the program is free.

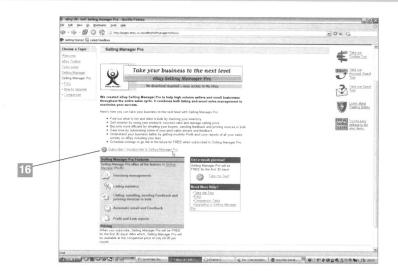

Selling Manager Pro

16 You may find that you'd like to upgrade to Selling Manager Pro, a tool that's especially useful if you're going to be selling in high volumes or have your own small business. Signing up to this service is free for the first 30-days, then £4.99 per month. Take the tour if you like, then click Subscribe/Unsubscribe to Selling Manager Pro.

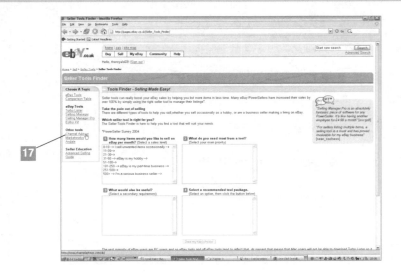

17 Once you've sorted all that out, you can see if any of the other tools are to your fancy – back at http://pages.ebay.co.uk/Seller_Tools_Finder, we click on Channel Advisor under Other tools.

For your information

So what are the main differences between Selling Manager and Selling Manager Pro? Well, apart from the cost upgrade you'll have to swallow if you go for the Pro tool, Selling Manager Pro also offers enhanced inventory management, listing statistics to help you analyse your selling processes, bulk emailing and sending-feedback capabilities, and profit and loss reports. This ability to monitor more closely the whole eBay sales cycle means that Pro is ideal for eBay shop and small-business owners. Casual but dedicated sellers should be fine with the standard version.

Serious business

18 Channel Advisor is another tool best suited to high-volume sellers. Its website at www.channeladvisor.com/uk has the info you need about what it does – the program you might be most interested in is ChannelAdvisorMerchant software, which helps you handle dozens of auctions at the same time, manage your inventories, promote yourself and ultimately achieve higher rewards for less work. This is really a tool to consider if your eBay selling is soaring away into the stratosphere.

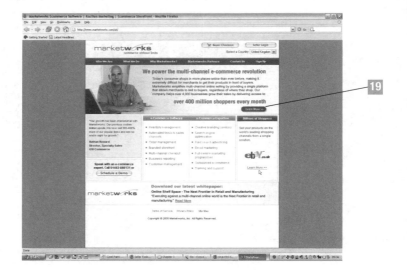

19 The page at http://pages.ebay.co.uk/Seller_Tools_Finder also mentions Marketworks (www.marketworks.com/uk), which again helps budding eCommerce entrepreneurs realise their dreams of setting up a thriving online shop. Marketworks has powerful tools which can help a store go from strength to strength – do your research on the services it provides via its website.

→ Other 3rd-party selling tools

Again, type 'eBay Selling tools' into Google and you'll get a wealth of links, some useless, some highly effective. eBay's partner PayPal (www.paypal.co.uk) offers a range of tools to help sellers – the range detailed at https://www.paypal.com/uk/cgi-bin/webscr?cmd=p/mer/directory_intro-outside, is pretty impressive (see chapter 5 for more). Keep your eye open for useful tools such as SitePal (www.sitepal.com), which lets you easily create an animated character for your website or auction listings, who can act as a virtual salesperson and really impress your jaded buyer. HammerTap, (www.hammertap.com) has: research software that lets you see just what sells on eBay and what doesn't; (Deep Analysis 2); a tool to help you calculate exactly how much you make on a sale (Fee Finder); and a suite of management tools to help you put your sales one step ahead of the competition (HammerTap Manager).

That's just the tip of the iceberg, so don't be afraid to do some serious research. If you reckon you're just going to be a causal seller, free tools and eBay's own free utilities are probably going to be enough for you, as they'll still put you a few steps ahead some of the lazier competition that's out there. Only consider huge, paid-for programs if you reckon you're going to get really serious with your eBay sales, and start selling at very high volumes.

Most of all, have fun and enjoy the buzz that selling on eBay can bring.

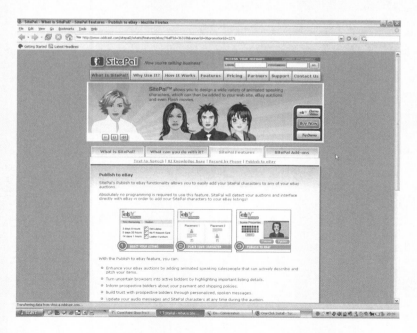

4 Build your own online store

WHAT YOU'LL DO

→ Browse through eBay shops

→ Read an eBay shops case study

→ Understand what makes a good eBay shop

→ Open an eBay shop

→ Benefit from more eBay shops tips

→ Open an online store with 3rd-party software

→ Find other 3rd-party software

One of the best options to explore on eBay, whether as a buyer or a seller, is eBay Shops (http://stores.ebay.co.uk/). As a buyer, Shops let you browse through similar kinds of items for sale from one seller – in the same way that you go to specific stores on the high street for certain items, so you can easily find eBay merchants who specialise in exactly what you're after. Many people who own eBay shops put a lot of time and effort into their store, are experienced eBayers, often PowerSellers, and know just how to keep their customers happy. For convenience, range of choice and speed of delivery, eBay Shops is well up there, competing with more conventional online stores.

And for sellers, of course, Shops offer an excellent way to build revenue and start making some serious cash out of the whole eBay experience. Setting one up need only take an hour or so, although if you want to compete with the best, you're going to have to realise that being a virtual shop owner is a full-time business, which needs hours of care and attention every week. Take your eye off the ball and there'll be hundreds of other shop owners ready to steal your thunder.

In this chapter, we'll dive into the world of eBay Shops, seeing what makes a good shop, what qualifications you need to get your own store off the ground, and how to build a shopfront from scratch. We'll also look at a third-party solution, Actinic Express, which lets you build a smart-looking store within minutes.

Don't forget, though, that building your store is just the start of an ever-evolving process. Maintaining your shop's stock levels, offering different buyer incentives, dealing with customers and understanding your market are just some of the things you'll need to be doing on a regular basis to try and make sure that once you get customers, you keep them coming back time and time again.

Pop along to http://stores.ebay.co.uk/ and you'll begin to see what an exciting off-shoot of eBay the Shops feature is. The Shops Directory on the left-hand side features dozens of categories that you know and love from normal eBay browsing – everything from antiques to sports memorabilia is there. You may be chomping at the bit to get your own store off the ground, but the first thing to do, as is so often the case on eBay, is to carry out some background research. Take a look at other shops, see what works and what doesn't, and start to get a feeling about how successful stores operate, and what the tricks of the trade are. An hour or so at this stage of proceedings will make all the difference when it comes to understanding how your own store can work, what it can sell and what the best tactics for success are.

First steps

1 You can go to eBay Shops from the home page at www.ebay.co.uk by clicking on the eBay Shops link on the left-hand side, or alternatively just go direct to http://stores.ebay.co.uk/).

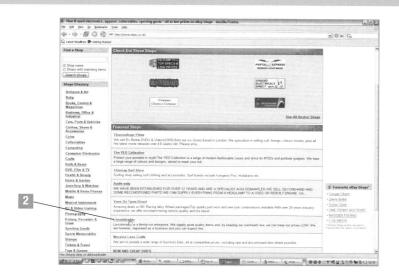

2 On the eBay Shops home page, you'll see a list of 'Featured Shops' to explore. Click on an underlined link to start your browse around what your fellow merchants have to offer.

3 As you can see, Shops are essentially just an extension of what you'll be used to in your normal searching through conventional listings. Shops should have some strong branding on their home page, with a headline and a logo to let you know just what you're looking at.

4 Obviously with shops you should find that there are numerous listings going on at the same time, normally in the same kind of field, although not always. The store we're looking at here (http:stores.ebay.co.uk/Lincolntrader) offers a wide range of items. Click on a listing that interests you.

Timesaver tip

Although it's very easy to set up an eBay Shop, as we'll show later, bear in mind that it's an incredibly competitive market, featuring dozens of PowerSellers and hundreds of people with acres of eBay experience. Unless you're a bit of a computer whizzkid and really fancy your chances from the off, we recommend that you spend a couple of months buying and selling through the main eBay channels, getting a feel of what works and what doesn't, before you jump into the Shops arena.

Research

5 Here's one of the items on sale from Lincolntrader, a deluxe gym exercise mat. You should be used to this kind of listing screen by now. As you would do when buying from an auction in the main area of eBay, do your research by clicking on the links under Seller Information.

6 This shop owner has PowerSeller status, which instantly lets you know that this is someone with a wealth of excellent eBay experience.

7 If you like what you see in a Shop, and reckon it could be a destination you'll want to return to again and again, on the main listings screen, click Add to My Favourite Shops.

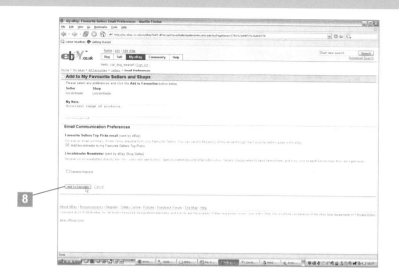

8 You'll be taken to this screen, where you can see some preferences for the addition of the shop to your list of Favourites. Add some notes about the shop if you like, before clicking on Add to Favourites.

For your information

What makes a good eBay shop? We'll be returning to this question a few times over the next few pages, but we'd say that our ideal store has strong branding, knows exactly what kind of thing it wants to sell, has new listings daily or as regularly as possible, has clear, well thought-out and thorough listings, excellent bargains, and customer service that's second to none. It can also be the case, of course, that a shop doesn't look anything special, but gains a special place in your heart because it specialises in selling a niche item that you love, and which is very hard to pick up anywhere else online. In these cases, what can seem a minor, innocuous shop can go straight to number one in your own list of favourite stores.

More tips

9 Go to the My eBay section of the site and click on Sellers under All Favourites on the left-hand side, and you'll see your newly added shop. Having a few shops in this list, built up over the weeks and months of using eBay, will save you lots of time and hassle when it comes to buying on eBay.

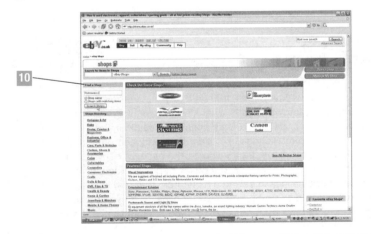

10 A friend may have recommended a shop on eBay, or you may want to do a general search for a shop name that is close to the kind of item you're after. In these cases, back at http://stores.ebay.co.uk, make use of the Find a Shop mini search engine in the top-left corner.

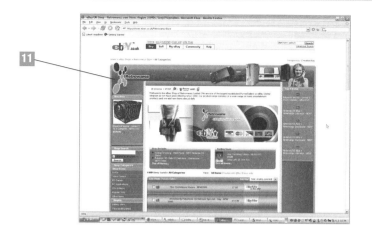

11 Our tip-off for Retrowarez takes us to this excellent store, which specialises in selling home entertainment products, such as video games and DVDs.

Handy searching hints

12 Use the Shops Directory on the left of the eBay Shops home page to narrow down your search and to find categories and sub-categories that you're specifically interested in. Clicking on Musical Instruments gives us a list of sub-categories to choose from, so we can be sure that the shop we hit on is heavily targeted at the kind of thing we're interested in.

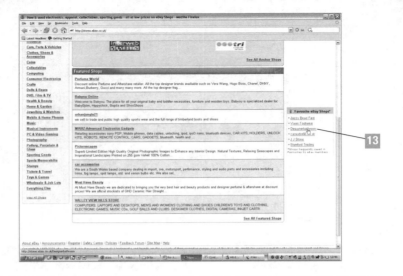

13 On the right-hand side of the eBay Shops home page you should also
be able to see a mini list of Favourite eBay Shops. This list is based on
the shops most frequently saved in Favourites lists by eBay members,
so every store featured should have something going for it. Click on a
link to explore further.

One of our own favourite eBay stores is Retrowarez (http://stores.ebay.co.uk/Retrowarez-Store). This store is one of eBay's largest shops, and has over 45,000 feedback ratings, over 99 per cent of which are positive. When we were online, the store had 13,739 items on sale, which is pretty remarkable. The store's boss, Steve Williams, has been an eBayer for six years, and the professionalism of the shop design and listings on the site instantly catches your eye. The company logo is prominently displayed on the site, and the range of Buy it Now offers means that if you're in a rush and out for a home entertainment bargain, this is one of the best stores on the whole of the internet to go to, let alone just eBay. The different sections of the site are easily navigable, and the How It All Began link gives you a real insight into the people behind the company, what makes them tick and what the future holds. The expansion that Retrowarez has undergone and its position near the top of the eBay Shops tree means that it's an ambitious sales model to try and emulate, but there's no harm in reaching for the top and being ambitious. If you start getting half of the loyal, satisfied customers that Retrowarez has achieved, you'll be doing extremely well.

Personal taste obviously comes into play here, and whether you regard aesthetics over content. A lovely looking store isn't going to guarantee sales, however – regularly updated stock, brilliant customer service and speedy delivery policies will help more in that regard. Don't be afraid to try and make your store stand out from the crowd, whether with niche items, incredible bargains, a touch of humour or just some elements of your own personality. Buyers will appreciate a touch of individuality behind your store, as long as it doesn't get in the way of the products that are on sale. If you can say truthfully that your stock is 'regularly updated' on your shop's home page, that will encourage people to come back again and again – as long as you stick to your promise and do, indeed, add new listings frequently. Think about refreshing the look and feel of your shop every now and again to keep things fresh, encourage customer feedback, and remember to have fun – becoming a Shopkeeper should be an enjoyable online hobby, not a chore. If it all starts feeling like a chore, then maybe eBay Shops is not for you, and you should stick to simple selling.

For your information

What kind of things can you sell in eBay shops?

The short answer to this question, of course, is 'anything'. In the same way that the main area of eBay caters for buyer and sellers of pretty much every item under the sun, eBay Shops has a cornucopia of the most diverse range of items you could possibly think of. Check out the eBay Directory on the left-hand side of http://stores.ebay.co.uk/ for proof of this, with the main categories broken down into dozens of sub-categories. Our own personal preference is for shops that specialise in certain items, such as home entertainment goods, video games or mobile phones – having one kind of item in your shop helps to keep things nice and focused. If you reckon your shop could be an Aladdin's Cave of all sorts of things, however, go ahead. The major thing to consider is that your shop can keep a healthy supply of new content coming – if you just sell a couple of things in it every so often, it kind of belittles the effort that you put into setting up a store in the first place. And remember, if you do have some niche items that you think would look good in a shop, don't be scared about listing them – you could end up with a smaller customer base, but also a more loyal group of people, who know that you're one of the few people to turn to when it comes to trying to find a certain obscure item or gift.

It's now time to set up our own store, using eBay's own Shop Builder step-by-step process. We won't look at listings, as how you list items has already been handled in chapter 3, but we will look at all the issues and techniques that surround getting your own shop online. To be in with the best chance of success, it will help if you've got a fair degree of eBay knowhow by now, and also that you have a very clear idea of why you're setting up store, and what it is that you think you can sell. It's best to think in the long-term when it comes to setting up shop – can you imagine selling the same kind of thing in a year's time? What does your shop offer that makes it stand out from the crowd? Can you afford the monthly fee and listings fees? Think about these questions carefully, and then dive in to one of the most fun and potentially money-making aspects of online life.

Get going

1 We've done our research and worked out what we want to sell, so let's get cracking. At http://stores.ebay.co.uk, click on Open a Shop.

2 If you see this screen, it means that you haven't satisfied the criteria to become an eBay shopkeeper. The criteria basically revolves around building your feedback up slightly, so go away and carry out a couple of transactions until your level is sufficiently high. With ratings of only 5 or 10 being required, this task should be a doddle.

3 Done that? Excellent. You'll see this screen. Click on Open Your eBay Shop.

Important

To set up an eBay Shop, you need to satisfy ONE of these criteria:

You are paying your eBay fees through Direct Debit.

You have a feedback rating of at least 10.

You have a feedback rating of at least 5, plus a PayPal account linked to your eBay account.

None of these demands is especially harsh – if you don't fancy Direct Debit, it really shouldn't take very long at all to build up the necessary feedback rating. Getting a PayPal account is almost something you do as a matter of course on eBay, so it's not exactly a hardship.

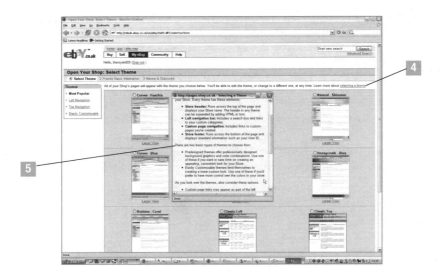

Name that theme

4 The first thing to do is to select a theme for your Shop, i.e. how it will look to the buyer. Options include Curves – Blue and Bubbles–Coral. Click Selecting a Theme to find out more.

5 Some useful tips to help you choose a theme pop up.

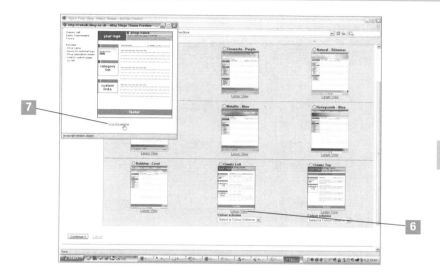

6 Click Larger View under a theme that catches your eye to see it in more detail.

7 A preview appears in the top-left, so you can see more clearly whether it's something you want to use. Click Close this window when done.

Timesaver tip

Why should you open an eBay shop? Well, pocketing more money is a major factor – normal sellers often see a jump of some 25 per cent in their sales once they've opened a shop (according to research detailed at http://pages.ebay.co.uk/storefronts/openbenefits.html#one). You can also pay less in listing fees by listing your items in the Shop Inventory format – you'll only have to pay 3p for each 30 days here. The lower price is because these listings will only appear in your Shop and in Shop Search, and won't be accessible in the main area of eBay.

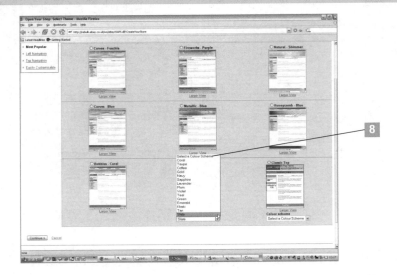

You've got the look

8 Then use the pull-down menu under your chosen theme to select a colour scheme.

9 Click in the circle next to your chosen theme to select it.

10 Then click Continue.

Timesaver tip

Opening a shop ultimately gives you increased credibility as a seller. You're effectively going from someone who just fairly randomly puts items on eBay, to a fully fledged virtual shopkeeper; if you're willing to spend the time and effort managing a shop and keeping content levels high, then it's a good indication to other eBayers that you're someone worth a look at. Of course, there'll still be bad eBay shopkeepers in the mix, but the time and effort that a shop demands weeds out a lot of the time-wasters who just want to get a quick buck out of eBay and run off as soon as possible.

The nitty-gritty

11 The next screen deals with the vital information for your shop – the name and description. A good name is vital for your shop's chances of success; see what eBay has to say about the matter by clicking on Naming Your Shop.

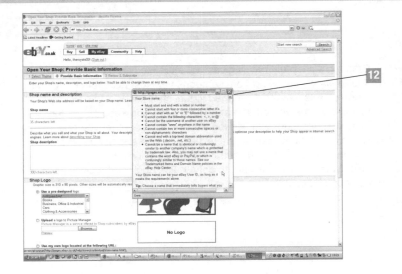

12 The list of criteria that your shop name must meet appears.

13 Do a bit of research back in eBay Shops about your name – does it already exist? We want to go for The Bookseller, and there aren't any shops with that exact title. Which is good news.

For your information

On completion of signing up for an eBay Shop, you get your own web address, based around the title of your store. This is another reason to become a shopkeeper – there's nothing like having your own URL to tell people about to inspire a bit of confidence and the belief that what you're doing is worthwhile and worth the effort.

Description, please

14 So we've got our title. Now on to the description of what your shop sells. Get tips in this vital area with the underlined link.

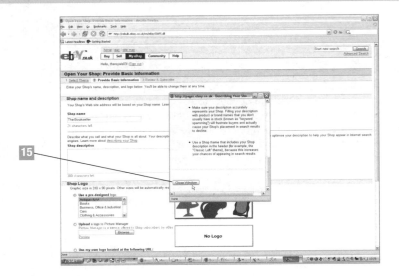

15 Click Close Window when you've digested the tips.

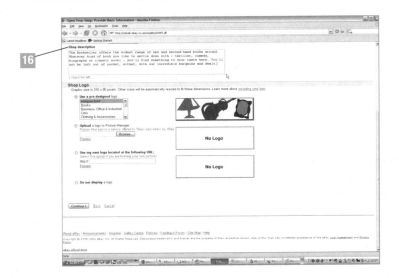

16 Then write your description. Watch the spelling and grammar, and stay to the point. You do have a limit of 250 characters, so you can't really waffle anyway.

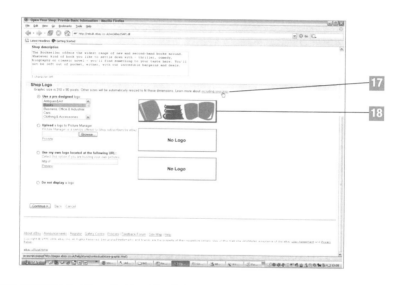

Lovely logos

17 The next step is the Shop Logo. Do you want to upload your own logo, or use eBay's pre-designed ones? Click Including Your Logo for tips.

18 eBay's 'Books' logo is shown here on the right.

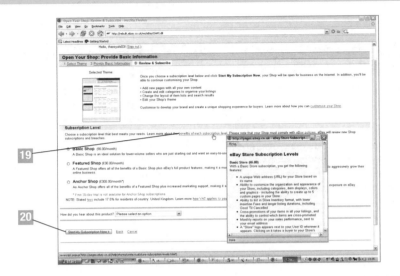

19 Moving on, we get to the Review & Subscribe page. Choosing your subscription level is necessary here – new shop owners will be looking at the Basic Shop charge of £6 per month, but click on Benefits of Each Subscription Level to get the lowdown.

20 We do indeed choose the Basic Shop option, and then click on Start My Subscription Now.

Timesaver tip

Seriously think about creating your shop's logo in a basic Paint program. As you can see in our example walkthrough, the logos that eBay provides aren't exactly scintillating, and something which shows you've put your own time and effort into branding will really impress potential buyers. You do need to think about how important aesthetics are in your shop, compared with content and customer service; in an ideal world, you should be able to combine both elements to your satisfaction.

We're online!

21 We can now open our shop! It even has its own URL, tied in with our shop name: http://stores.ebay.co.uk/The-Bookseller. Click on the link to go to it online.

22 It's all a bit sparse and empty at the moment, and could certainly be shouting about itself a bit more. The logo is also a bit garish, although you can't really tell that in this black-and-white screenshot. It's very basic and we could liven things up a bit – it's important to be hyper-critical about your shop, because if you don't like it, no one else will!

Manage it

23 Now that you've had a sneak peek at what the store looks like online,
it's time to do some editing and tweaking. Customising your shop with
new pages, new content, new categories and different layouts helps to
keep things fresh. Click on Go to Manage My Shop.

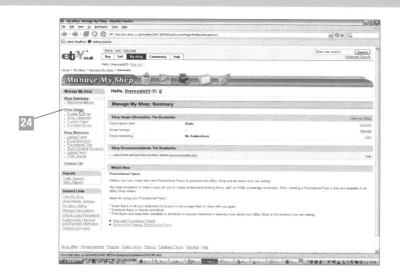

24 A summary of what's going on appears in our old friend, My eBay. You can tweak to your heart's content here, and really micro-manage your store. Click on Shop Design to start changing things.

Timesaver tip

Whatever kind of shop you have, and whatever you're selling, there's one golden rule which we make no apology for repeating – the customer comes first! If you're emailed with sensible queries from buyers, do your level best to answer them as quickly as possible – you don't have to be standing by your PC every minute of the day, of course, but being able to devote half an hour or so whenever you can, just to deal with questions, could make all the difference when it comes to the success of your shop. Answer people promptly and courteously, and then deal with them thoroughly and professionally should they win an item, and you'll be well on your way to impressing the eBay community with your shop credentials.

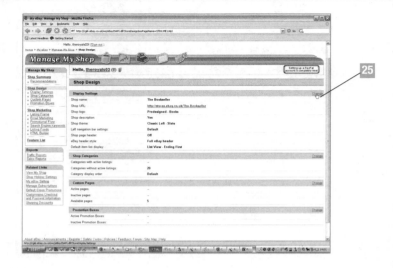

Time for change

25 To change any of the settings, click on Change on the far-right.

26 Maybe you don't like the theme, and want to see how items could look if they were presented differently. If so, click on Edit Current Theme.

Timesaver tip

Of course, just because you have an eBay Shop, that doesn't mean that you should discount your other methods of selling. Try and strike a healthy balance between your normal auction listings and your shop sales. Why not promote your shop in your auction listings? If people like your one-off sales, they might be tempted to go to your store and see everything that you have to offer.

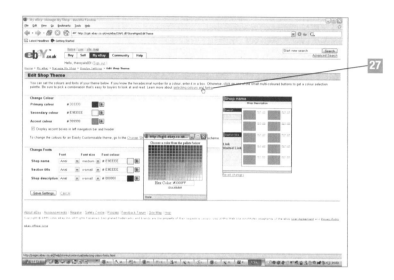

27

Tweak it

27 You can play around with colours and fonts to your heart's content. Click on Learn More About Selecting Colours and Fonts to see what tools are on offer to you.

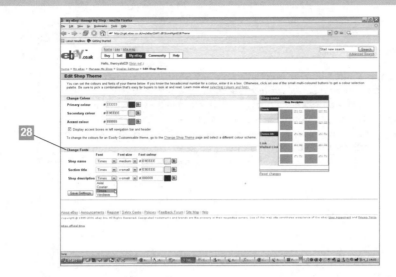

28 Use the pull-down menus under Change Fonts to play around with font types.

Timesaver tip

Setting up your shop is only the beginning, as we've mentioned. You need to start spreading the word about your store so that as many people as possible can see what you have to offer. As well as promoting your shop in your normal auctions, get friends and family to start spreading the message – virtually everyone knows someone who's on eBay, so there should be no excuse for people not knowing that you're up and running. Put your shop's URL in your email signature, so whenever you correspond with a fellow eBayer in a transaction, they can see that you have a store. You could even think about creating business cards to send out to customers as well.

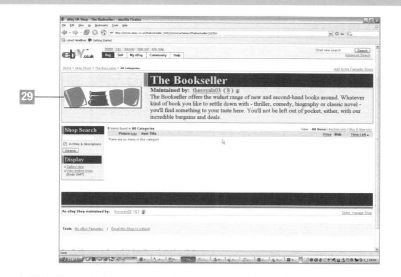

No logo

29 The store is shouting about itself a bit more now, although the logo isn't doing us any favours. Be critical – if it looks horrible, then bin it! We decide to see what the shop looks like without it.

30 Without a logo, it's a bit tidier, but rather too dull and text-heavy. Obviously, adding our listings shortly will liven things up, but we're going to go away and create a little logo in our PC's paint program, to add a touch of individuality and class to proceedings. Little touches like this can really help sell your shop – why keep something on there that you don't like? You're in charge, so start making some tough decisions.

Timesaver tip

If it all goes wrong and you do find that a shopkeeper's life is not for you, you can close down your shop through the ubiquitous My eBay section of eBay. You'll get an email from eBay confirming the closure, and whether there are any unpaid fees or not. You can re-open or set up a new shop pretty easily, so it doesn't necessarily have to mean the end of the world if you take this course of action. If things aren't working out, it may be best to close and re-think your action plan, rather than struggle for ages getting nowhere.

Get listing

31 When you come to list an item now, a third option will appear – Sell in Shop Inventory. Choosing this option means only eBay Shops-visitors will see your item, but it does cut down on your fees.

32 Then set about creating some stylish listings to populate your shop and stop it looking so bare. A dozen starting listings or so will raise people's interest, and as long as you have more stock to put online over the next few weeks, you're now well on your way to having a thriving shop. Right, we're off to get a decent logo sorted! Remember that eBay Shops is a serious commitment, so be prepared to tweak, edit, change and react to market conditions at very short notice. Stay on top of the game and success should come your way!

→ More eBay shops tips

What we've talked about here is only the tip of the iceberg – the hard work comes in the next few weeks, months and even years. Spend time and effort sorting out your shop to your satisfaction – don't expect it all to slot into place in the first week. Look at different methods of promotion, including creating your own promotional flyers to let people know about you and what you have for sale. Go on the eBay Stores Discussion Board to get a feel about what your fellow shopkeepers are up to – you could find some useful nuggets of information here, although it's unlikely that anyone will give away their closely guarded auction secrets. Don't be disheartened if trade is slow at first – ask yourself, is your shop not working the way you want it to? Could you experiment with different layouts, even different product lines, or have a week of rock bottom prices to encourage trade? Do you have an online rival who's offering the same kind of products at lower prices? The learning curve is steep and, obviously, people with huge feedback scores and years of experience will have a head start on you – but if you bide your time, don't expect miracles, and concentrate on offering something people actually want, at low prices with thorough customer service, you should find things pick up, and people start to come to you for a first-class eBay shopping experience.

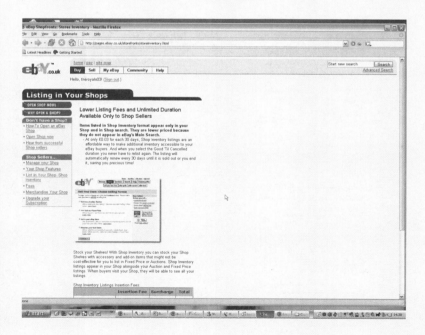

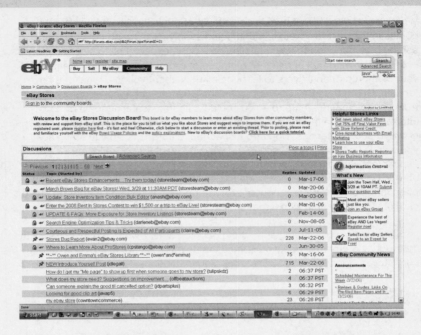

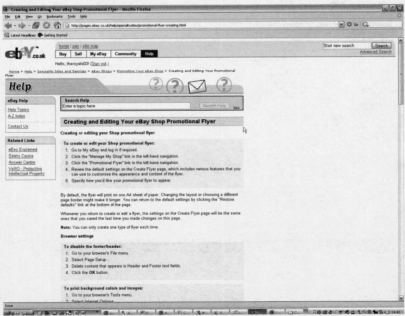

Obviously, eBay would like you to use its own service when you open up a shop online. Indeed, it uses the page at http://pages.ebay.co.uk/storefronts/comparisons.html to pinpoint the reasons why it thinks you should do this – the main reason being that millions of people visit eBay in the first place, almost guaranteeing a well-designed shop plenty of visitors from day one. Use a third-party service and you may struggle to get people to know that you actually exist.

However, this is a simplistic argument, and we're duty bound to show you how to build an online store with a powerful, cost-effective and stylish alternative to eBay – Actinic Express. Actinic (www.actinic.co.uk) has been around for nearly ten years now, providing eCommerce solutions to countless small and medium-sized firms, primarily through its Actinic Catalog software. The new Express option is a low-cost, very easy to use piece of shopping cart software that contains everything you need to get a professional-looking web store online, with its order-processing facility enabling you to start selling pretty damn quickly as well. The online demo lets you understand how the program's features work, and we'll talk you through that over the next few pages, look at how an example shop, created using the software, operates.

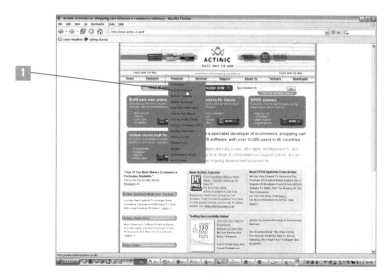

Find the demo

1 Here we are at Actinic's UK home page at www.actinic.co.uk. Have a little nose around the site if you like, then click on the Products tab at the top of the screen, followed by Actinic Express.

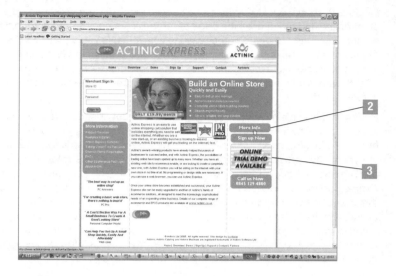

2 Actinic says that its Express software is ideal whether you're a new start-up, or an existing business looking to expand. A main attraction from our point of view is that it really is brilliantly simple, yet flexible to use – and you don't need any complicated technical knowhow, which is where so many other 3rd-party solutions fall down. Click More Info.

3 The best way to get a feel for the program is to take the online trial demo, at absolutely no cost to yourself. That way, you can see whether the program is likely to be something that you wish to invest in – presumably you're already interested in viable alternatives to eBay's Shop option, but it's certainly worth being sure. Click Online Trial Demo Available.

For your information

Obviously, playing around with the online trial demo won't set you back a penny, but seeing as the demo is available to all users, if you're a fan of the software you're going to want to sign up for it permanently. To do this, call the Actinic Sales team on 0845 129 4800, and expect to pay £49.99 for setup and £19.99 per month (+ VAT). Features you get for your money include web hosting, a web site (obviously), product catalogue and shopping cart, order-processing functionality and telephone and email support. More details about signing up are at www.actinicexpress.co.uk/sign-up/index.htm.

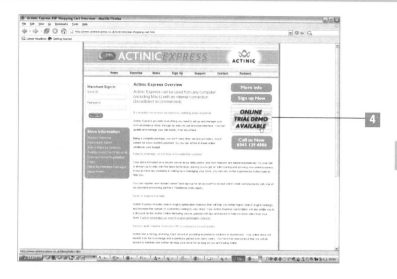

Startup Wizard

4 There's one final screen of blurb. Click on the Demo link again.

5 The demo loads up. The major points of the demo are to show you how the admin software works, and how customers will be able to buy your products. You can even add and edit categories on the demo, although they'll revert to default values every two hours. On the right-hand side of the first demo screen, click Startup Wizard to start setting up your store.

For your information

Actinic Express is by no means the only product from the company. Click on Products from the home page at www.actinic.co.uk to see some other solutions, with perhaps the most well-known being Actinic Catalog, a shopping cart software package that has now reached Version 7. Again, there's a free 30-day trial version for you to get to grips with the software. If you decide to purchase, it will set you back RRP £379 + VAT. Obviously you're paying far more here than you would do with an eBay shop, but the features on offer tower above the basic options available to eBay users. If drag-and-drop editing, unlimited sections, fully customisable design, comprehensive layout and colour options, and in-built powerful marketing features sound good to you, you may find the asking price an absolute steal.

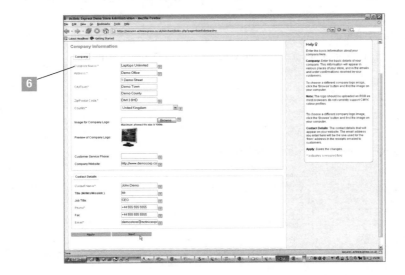

Lay down the law

6 Before your store can take orders, company information, terms and conditions, tax, shipping, payment and credit card settings need to be sorted out. Here we see fields needing to be filled in with your company information.

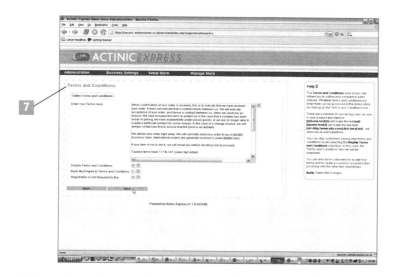

7 Terms and Conditions are next, letting your customers know important site policies.

Timesaver tip

It should really go without saying, but make sure you get all the admin details of your site exactly right, with clear policies on important issues such as shipping and terms and conditions. Your potential shop visitors need to know exactly where they stand when they visit your shop, and any vagueness or lack of clarification will work against you in the long run.

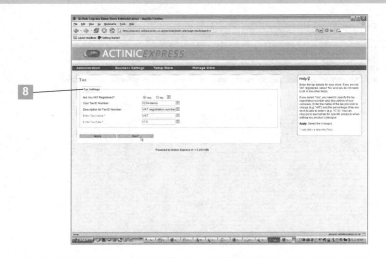

Taxing questions

8 The next screen holds the VAT and Tax questions.

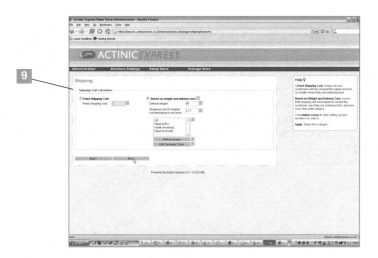

9 Then you need to sort out shipping costs. Note the help boxes that pop up on the right-hand side to give you further assistance.

Timesaver tip

Take a look at some stores that have been set up using Actinic
software, to get a grasp of how versatile it really is. Examples include
Premium Tyres Online (www.premiumtyresonline.co.uk), Coolcards
(www.coolcards.co.uk) and the gifts, gadgets and gizmos shop
Tiggypig (www.tiggypig.com). If these people can set up successful
shops, why can't you?

4

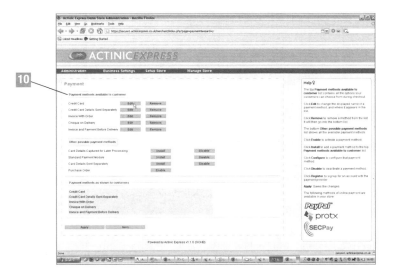

Perfect payment

10 Vital payment questions crop up next – what methods do you want to
be available to customers? Edit, Remove or Install are your options,
with PayPal, Protx and SecPay available as secure online payment
choices. If you click on Edit you'll be taken to a further screen.

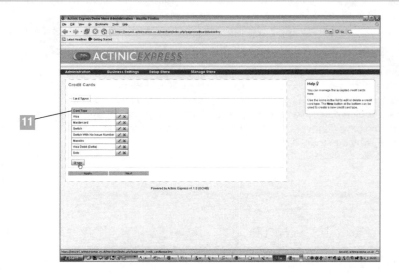

11 Here, we're looking at the different options for credit cards.

For your information

Actinic has a comprehensive Knowledge Base (http://knowledge.actinic.com/acatalog/) and Technical Support home page (http://www.actinic.co.uk/support/index.htm) which you can turn to in times of trouble. You can request help through the telephone hotline by calling 0845 129 4888.

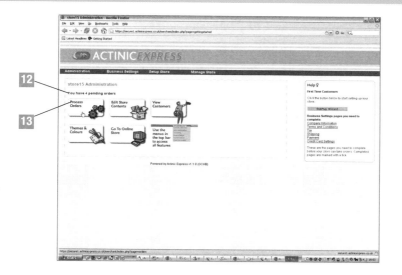

Let's manage

12 Over the next few pages, we'll look at an example shop in action. On the left-hand side of the online demo, you can see numerous icons to help you manage your store – process orders, edit store contents, view customers, themes and colours, and go to online store. We're told we have 4 pending orders.

13 We click on Process Orders.

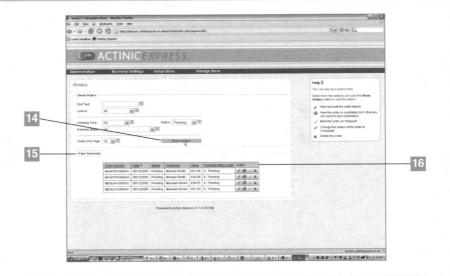

14 Click Show Orders to see the details of pending orders.

15 The Order Summary shows you exactly what orders have been placed and when, their status, the customer name and the value of the transaction.

16 Click on the icons in the Action box to start dealing with the orders.

For your information

The Actinic Download Centre is a great place to go for the latest trials, software patches and updates, help links, plug-ins and PDF user guides. Go to http://www.actinic.co.uk/support/downloads.htm and see what takes your fancy.

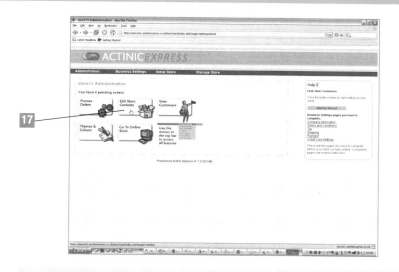

Content is king

17 Now to deal with the content in our shop. Click Edit Store Contents.

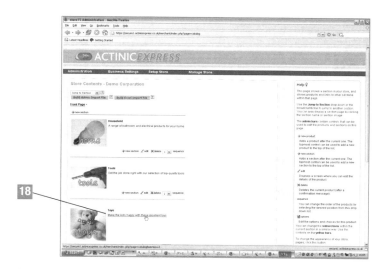

18 Thumbnail pictures of the kind of things that your shop sells appear. Click on one of the descriptions to explore further – under Toys, we click on Make the Kids Happy With These Excellent Toys.

Did you know?

Actinic has the backing of some well-known companies in the UK, who use the service for their own eCommerce needs. Huge UK clients include the Royal Opera House, Land of Leather, The British Library and Oxford University. The fact that Actinic has such high-profile users is a good indication of how highly its services are regarded.

Product range

19 Here you can see the toys on offer, their price and a link to add them to a shopping cart. You can jump to another section of your store easily with the pull-down menu in the top-left.

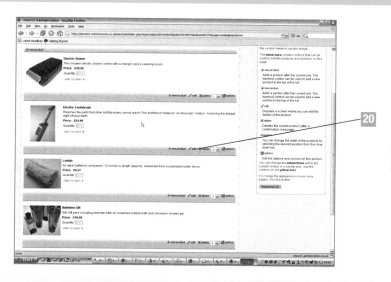

20 We click on Bathroom to see the range of bathroom products. Don't forget the New Product, Edit and Delete options towards the right-hand side, which let you control just what products you have on offer.

Timesaver tip

Setting up a store online can be a scary business, so it helps to read about like-minded users who've been through the process, used the software and come out trumps. Actinic is understandably keen to show off real-life examples of firms who've used its software to positive effect, and reading the case studies at www.actinic.co.uk/examples/case.htm does reinforce the fact that the company's software is genuinely helping hundreds of small businesses make it big online.

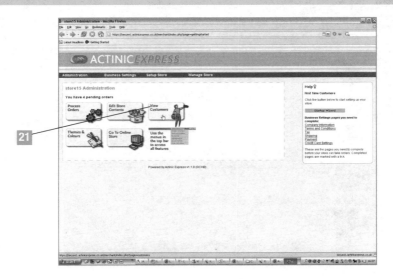

Customers

21 Back on the main admin screen, click View Customers.

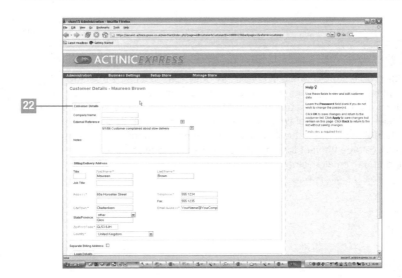

22 Details of your loyal customers pop up. Again, full Help links are on the right.

Important

Remember, you can add as many products and categories to your shop as you like – there is no limit. Having a bulging product catalogue is one way of impressing casual visitors and making them understand that you really mean business.

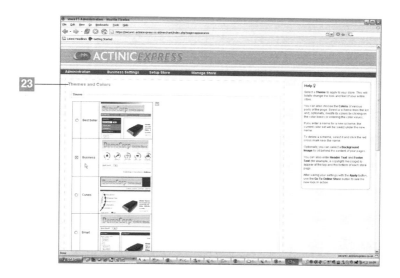

Themes and colours

23 Are you happy with how your store looks? Contemplate a lick of paint, or a radical overhaul, by clicking on Themes & Colors from the main admin screen.

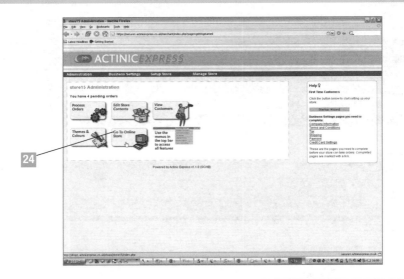

24 Click Go to Online Store to see it all live on the net.

Timesaver tip

Obviously, the design of your shop is important to its overall success. You'll see in these Actinic screenshots that everything is kept quite clean and classy, with the customer not being too over-burdened with conflicting product or pricing details. Make sure the design of your site reflects the products that you're selling – you'd be surprised how many shop creators build something that looks totally inappropriate and unconnected to the actual items that are on offer. Remember that Actinic Express requires no real design or programming skills, so choosing the templates wisely, and adding your own photos, logos and colours, is where you really need to excel.

Shopping cart

25 And here is the online store! Click on a link to check everything is as it should be.

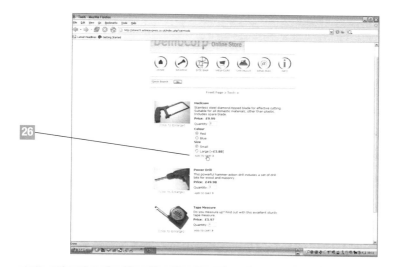

26 The shopping cart experience is all-important. Customers need to be able to buy quickly and efficiently, with prices, sizes, colours and any other details spelled out.

Timesaver tip

You can reward regular users of your Actinic shop with customer accounts, to encourage loyalty and get them coming back again and again to your store.

For your information

One of the most important components for a eBay rival is to have in-built marketing features, to help your shop get noticed. After all, eBay has the benefit of millions of users coming to its auctions every day, so a potential rival really needs to offer something good as an alternative. Actinic Express has automated search-engine optimisation features, which work hard to ensure that your site gets high search-engine rankings, and thus gets seen by as many people as possible.

27 Customers need to be able to view their cart at any stage of the shopping process, so they can have a running total of what they're going to be spending.

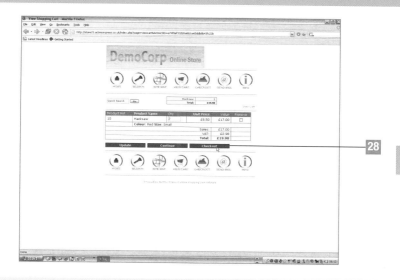

28 Charges are laid out clearly and effectively, so the customer can go to the checkout without fearing any nasty surprises. Build your shop according to how you'd like a shopping experience to be, keeping the customer's convenience and preferences at the forefront of proceedings at all times.

→ Other 3rd-party software on the market

We can't leave this area of eBay and online life without mentioning that there are, of course, many alternatives to Actinic when it comes to building an online store. We chose Actinic Express for its powerful features, low costs and sheer ease of use and flexibility; some of you may like to trawl your net wider and try other alternatives. EROL (www.erolonline.co.uk) has won plenty of fans for its powerful features, one-off payment fee and friendliness of use; you can download a free trial of EROL 4 Business at http://www.erolonline.co.uk/download_trial.asp?section=products. ShopFactory, meanwhile (www.shopfactory.com), has been responsible for over 150,000 online shops since it started over ten years ago, and has plenty of converts to its powerful software range. Click on the Free Demo link on the home page to find out more.

These are just two alternatives, but combined with Actinic they show that eBay Shops is by no means the only option for would-be virtual shopkeepers. There's little doubt that eBay has a massive advantage with its pre-installed customer base with millions of auction fanatics, but these rivals have hit back with powerful features and some classy design which means that you're never left scratching your head wondering what to do. And with the free demos available, you've got nothing to lose by testing the rivals out, and coming to your own informed opinion about what's right for you.

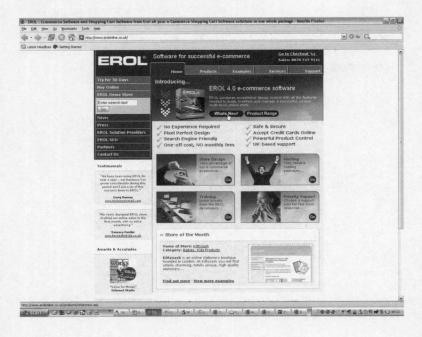

Timesaver tip

eCommerce tips

Hopefully you'll have grasped some important tips about shop building over this chapter. The phrase 'eCommerce' is unfortunate, really, as it tends to conjure images of massive, technically oriented web businesses, rather than simple yet highly effective personal stores. Type 'eCommerce tips' into Google and you'll get primarily business-led tips. Some basic facts should hold you in good stead: know what you want to sell, spend time beforehand planning, make your store useable and accessible, pay attention to payment security, offer excellent delivery options and customer service, look at ways of promoting your store online and off, keep your prices competitive, know your market, offer help when things go wrong, lay out a clear set of terms and conditions, and react quickly to sudden fluctuations in the market. Enjoy yourself and don't hide your light under a bushel – it's a competitive world out there! Some useful small business advice can be found online at www.bytestart.co.uk.

5 Safe and sound online payment

There's no getting away from it – if you're going to become a successful eBayer, whether in the fields of buying or selling, sooner or later you're going to have to dive into the world of online payments.

Thanks to numerous scare stories in the media over the last few years, detailing horrific tales of people losing thousands of pounds online and having their accounts cleaned out, this part of the internet experience is still one which fills many people with a certain amount of apprehension.

The good news is that there's no need to be worried, as long as you stick to the major online firms that deal with secure internet payment. And when it comes to eBay, you can hardly fail to notice that one company has a massive market domination in place – PayPal.

PayPal (www.paypal.com) has risen from relatively humble beginnings to become a major player in online finance. And if you were wondering about why it's managed to rise so highly in such a short space of time, there's one simple answer – it was bought by eBay in October 2002, to become one of the auction site's growing list of partners. So every time you visit eBay's home page, you'll see, amongst all the auction links, plenty of advertising and promotion for PayPal. There's even an eBay area dedicated to PayPal, at http://pages.ebay.co.uk/paypal/index.html, where the service is described as a way to 'send and receive online payments safely, easily and quickly'.

eBay wouldn't have gobbled up PayPal if it wasn't sure of what it was doing, however, and PayPal's success has also been built on its simplicity, and the fact that it's free, with no set-up fees or monthly charges to worry about. For buyers, you can pay through PayPal instantly, and let it worry about the credit card details and suchlike – the seller will never see your details, thus eliminating a huge chunk of most people's fears about online trading. It's worth repeating that – your financial information is NEVER shared with the seller.

And for sellers, you get paid the monies instantly into your account, and can then either choose to transfer the money into your bank account, or leave it as credit online, ready to use the next time you need it.

In this chapter, we're going to assume the mantle of first-time visitors to the PayPal site, signing up and sorting out our account with the company. We're also going to look at one of PayPal's rivals in the market, the well-respected Nochex, just to show that you're not limited to one payment aid – you have a choice.

You can sign up with PayPal through the numerous links on the eBay site itself, but going to the payment firm's UK home page at http://www.paypal.com/uk/ is a useful exercise, as you can see the whole range of services that are on offer. Registering with PayPal, which we're going to do below, is very easy, but there are a few things to consider, such as what kind of account you want. Tailoring the type of account to what kind of user you are is very important, and will help you get the most out of PayPal. Let's see what the service has to offer, over the next 8 pages.

Take a trip around

1 How you want to tour the PayPal site is up to you, obviously, but we're going to have a look around whilst signing up. Click on Sign Up Now from the home page at www.paypal.com/uk.

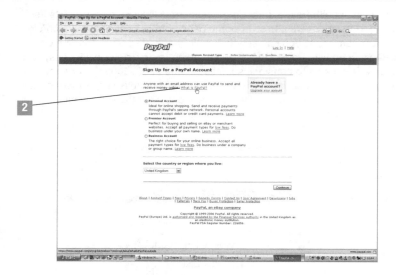

2 As we know, PayPal is a simple way to send and receive money online. We're up for doing a bit more background digging, however, so click on What is PayPal?

Did you know?

PayPal has over 88.6 million accounts worldwide, full of people doubtless attracted by the safety and ease of use of the service.

Did you know?

PayPal's secure global network encompasses more than 50 countries, and if you live in Australia, Austria, Belgium, Canada, China, France, Germany, Italy, the Netherlands, Spain, Switzerland, the UK or the USA, you can access sites specifically tailored to your own region.

Important

Never send cash to a seller for an item you've bought on eBay – it's simply not worth the risk.

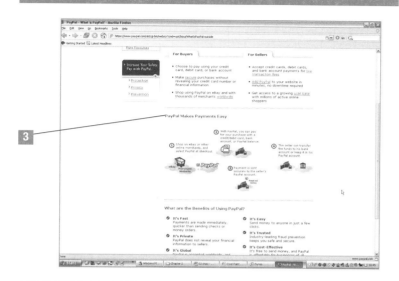

Find out more

3 The advantages for buyers and sellers are clearly explained on the next screen, with a diagram under PayPal Makes Payments Easy giving you a step-by-step rundown of how everything works – from you shopping on eBay to the seller getting his or her money. The fact that this process can be completely summed up in four short steps shows you just how simple the PayPal scheme is.

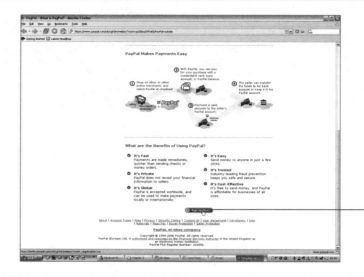

4   If you've read all that and digested all the benefits of the service – as we've mentioned, the key plus point really is the fact that PayPal ensures that none of your financial information is revealed to sellers – and are happy to continue, click again on Sign Up Now!

Timesaver tip

As should be expected with a service as popular and heavily subscribed to as PayPal's, there's an excellent Help service accessible from the Help tab on the home page. Questions are divided up into several categories on the left-hand side of the Help page, including making payments, tracking payments, withdrawing funds and adding funds. Simply click on a folder icon to dive into the Q&As and helpful sub-categories.

Did you know?

eBay claims that fewer than 0.01 per cent of eBay transactions end in a case of confirmed fraud – pretty impressive, although that still amounts to some 300 cases every day in the UK alone.

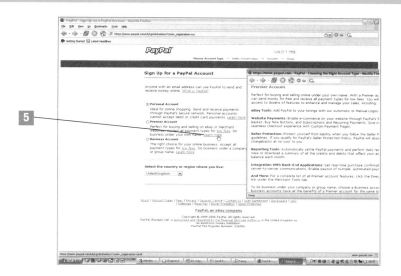

First steps in registering

5 Key to your time on PayPal is deciding which kind of account you want
– personal, premier or business. See our summary below to find out
what kind of account suits you best, or simply click on the Learn More
link under each category to open up a new window on the right-hand
side of the screen, telling you everything you need to know.

6 Make your decision based on your knowledge of what kind of user you're likely to be, clicking in the little circle next to one of the three headings. We go for a Personal Account, confirm that we live in the UK in the pull-down menu below, and then click on Continue.

For your information

When you sign up for PayPal, you'll be asked whether you want a personal account, a premier account, or a business account. A personal account will be fine if you reckon you're primarily going to be a buyer online, as it lets you send and receive money for free – but not accept debit or credit card payments. This caveat leads us onto the Premier Account, which is the one for you if you're a buyer and a seller, as it lets you send money for free and receive all the different types of payment, for a minimal fee. The Premier Account also has a host of other features useful to a seller, which are detailed on the PayPal site.

If you're a serious user who has their own online business or shop, you'll want to think about the Business Account, which has technical tools that will smooth the process of selling your wares to a potential audience of millions.

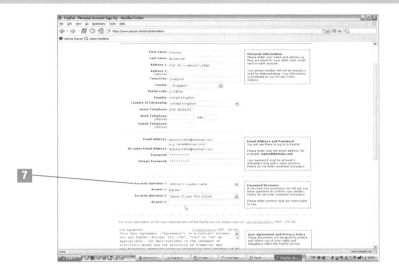

7

Personal info time

7 Now it's time to enter some simple details, such as your name, address, email address and password. Include some numbers in your password to heighten its security, although make sure that it's a password that you're likely to remember. You'll be given a couple of security questions at the bottom of the screen, which you can set PayPal to ask if there are any problems accessing your account.

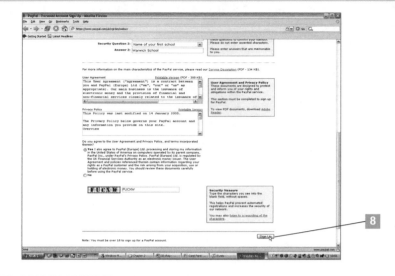

8 Once you've sorted out the security questions and answers, read through the user agreement and privacy policy, tick the buttons at the bottom of the screen, enter the characters you see in the space provided (this is an added security measure, to help PayPal stop automated registrations and the threat of fraud) and click Sign Up.

For your information

The personal details you give PayPal are protected by the site's Privacy Policy, which is detailed in full at the bottom of the sign-up screen.

Important

Try and make your password as unique to you as possible, but still memorable. It is case sensitive and must be 8 characters long. It isn't a good idea to have one password for dozens of different websites – many online surveys have revealed that this is a tactic which many people in the UK employ. It may aid your memory, but it also increases slightly the chances of your online security being compromised. Unfortunately, we're now living in a world dominated by countless pin numbers, passwords and usernames, but it's worth sharpening your memory up and having some different details for different sites, to be on the safe side.

5

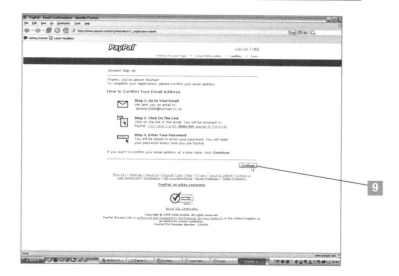

Email Confirmation

9 That step completed, you now need to confirm your email address. PayPal will send an email to the address specified in your signing up procedure. In the email will be a link which will enable the site to complete your registration and confirm the validity of your details. If you wish, you can choose to confirm your email address at a later date by clicking Continue – you may be in a rush and just want to finish up as soon as possible without accessing your email.

10 If you do have time, though, it's best to go to your email account and get the confirmation out of the way. We go to Hotmail and sure enough, there's an email from PayPal, which we click to get our unique link and confirmation number.

Important

Because of the way that some email programs work, there is a chance that the link you're sent in the email from PayPal won't work. If this happens, don't panic – go back to the email and note down the confirmation number that should be there as well. Then go back to PayPal, log in to your account and click the Confirm eMail link in the Activate Account box.

A couple of options

11 Once your email address has been safely confirmed, you have a couple of options. You can either go back to your email account, where you should have a handy message from PayPal again, this time giving you some excellent tips on getting started with your account.

12 Or, you can log in at PayPal, and click on My Account to see your personal account overview. You'll see that your account is unverified. This is because PayPal initially puts a limit on how much you can send and withdraw, for the purposes of security. To lift these limits, you need to become verified. The verification process varies from user to user and depends on your account status, but basically it is likely to involve you setting up bank funding, by giving PayPal your bank details, and then telling PayPal the value of two small security amounts which it will have dropped into your bank account. If you pass this security test then you'll be able to start sending PayPal payments that are funded by your own bank account.

Timesaver tip

Is it worth getting verified? You may be apprehensive about giving your bank details to PayPal, but don't forget the site's massively strong security record – everything will be safe. And if you do get verified, it may impress other online users who are thinking of trading with you – it shows that you're trustworthy and have nothing to hide.

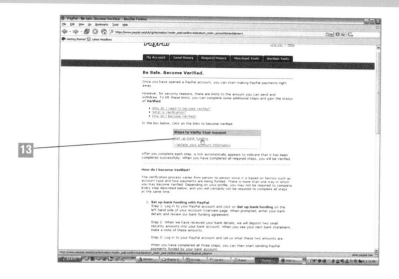

Verification

13 If you decide to go for verification, you'll be taken to this screen, which details a little bit more about the whole process, telling you what you'll need and what you have to do. Read through the blurb and then click Set Up Bank Funding.

Timesaver tip

The My Account part of PayPal plays a similar role to that of My eBay in eBay, acting as a convenient home for all your important transaction details. Click back here regularly to see a summary of all your PayPal activity, the latest site news and your profile. You can also add funds and withdraw monies from here, as well as access the Resolution Centre, which lets you get to the bottom of any disputes, or file a complaint.

14 Then just go through the steps of adding your bank account – here, we're confirming that it's a UK bank account. We won't go through this in detail as everyone's experience will be different, but the step-by-step process should mean that it only takes a couple of minutes, and everything's totally secure. One way you can tell this, aside from the padlock that should be in the bottom of your browser window, is by looking at the address in the address bar at the top of your browser – it should start with 'https://', the 's' being the indicator of security.

15 Click Continue to go through the steps of bank verification.

→ PayPal extras

As befits a partner of eBay's, PayPal has plenty of tools for you to get your teeth into, once you've successfully sorted out all your registration details and tasks. Remember that the very fact that you've signed up with PayPal shows that you're a serious eBayer – people who buy on eBay are more likely to buy from you if they see that you're a fully sorted out member of such a secure, convenient way of handling online money transfers. The two areas of PayPal you'll want to explore are Merchant Tools and Auction Tools. The auction tools are a boon for sellers – for example, the Automatic Logo Insertion facility (https://www.paypal.com/uk/cgi-bin/webscr?cmd=xpt/auctions/Automatic LogoInsertionLanding) adds the PayPal logo to all your listings, saving you the time and hassle of having to add it to each individual sale you decide to put up on the site. The protection programmes which are offered to both buyer and sellers are a welcome feature as well, giving you that added peace of mind that, if things go wrong, you have something to fall back on to help you sort out the problems. If you want to take a look at all the tools PayPal offers, and whether they're suitable for your own unique needs, take a trip to the summary page at https://www.paypal.com/uk/cgi-bin/webscr?cmd=p/mer/ directory_intro. Many of these tools will only be useful to you once you've built up a good deal of experience on eBay, but just knowing that they're around is yet another plus point in the PayPal experience.

5

The likelihood is that PayPal will be perfect for all your eBay needs, and that its track record in safe and convenient online payments will convince you that it's the best service around. However, to say that it has the market totally sewn up, or that it is the only option around, would be misleading. Viable alternatives to PayPal include Protx (www.protx.com), WorldPay (www.worldpay.co.uk), the eBay-recommended Escrow (www.escrow.com), and the site we're going to look at below, Nochex (www.nochex.co.uk). Nochex has a number of supporters online, and is especially suitable if you've got your own online shop – its home page blurb states that it 'provides a secure, reliable online payment service to thousands of small and medium sized businesses like yours'.

We'll have a scout around the service below.

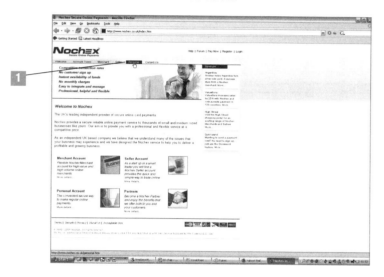

First steps

1 In a similar fashion to PayPal, Nochex offers a number of different accounts for you to get your teeth into – there's a Merchant Account (high value, high volume sellers), Seller Account (good for eCommerce startup businesses) and a Personal Account (the standard way to make regular online payments, similar to PayPal's Personal offering). Explore the options as you wish. We click on Personal.

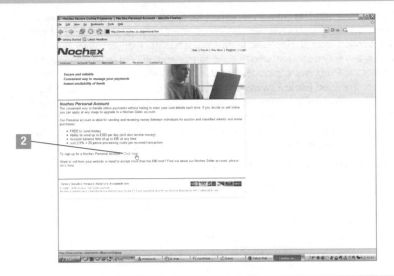

2 Some of the benefits of a personal account appear on screen – it's free to send money (which we think is only right), and you can send up to £300 per day, as well as receive money. It's the one for us, so we click on the link next to Sign Up for a Nochex Personal Account.

Timesaver tip

If you want to look at some Nochex success stories, take a peek at http://www.nochex.com/casestudy-argentice.htm. Case studies are often used by payment providers such as Nochex, helping to give the interested punter a real-life slant on proceedings, and see how business can be rapidly improved just through judicious use of the right service.

For your information

The Nochex success story we saw when we visited the site was based on the Argentice website, at http://www.argentice.co.uk/shop/.

Details, please

3 More details of the Personal Account appear. Simply click Continue.

4 Personal account registration time. Enter your name, email address details, password and password hint. Click Continue again when done.

For your information

You can discuss Nochex issues, find out the latest updates, read success stories, make complaints or just engage in some good old fashioned auction banter in the forums, situated at http://greenroom. nochex.com.

For your information

Perhaps the most important thread in the forums is Newbie Questions, under Self Help. Come here first before asking loads of questions, as many of the answers you're after are probably in here. The thread Dos and Don'ts – What You Can and Cannot Do With Nochex is a very good place to start.

Check your email

5 Just as with PayPal, you'll now be asked to check your email. This time around, a 4-digit confirmation PIN will be waiting for you.

5

PIN please

6 At the top of the Inbox is your message from Nochex – click on it to get the pin number contained within.

Timesaver tip

There's an excellent 'Quicksend' facility on Nochex which lets you send a payment quickly, with no need to sign up or anything. You enter the email address of the person you want to send the money to, fill in the amount, type an optional message, then press Continue and follow the instructions. There is a caveat that if you don't have a Nochex account, depending on who you are sending money to, you may be asked to sign up for a full account. Nevertheless this is a handy option to have.

Take a look at https://www.nochex.com/quicksend/ for more info.

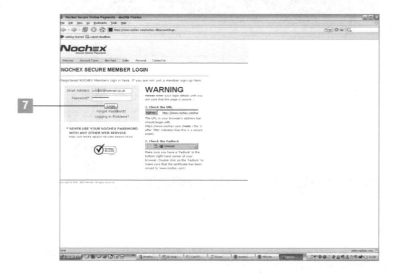

Get it right

7 Now sign in through the secure login page, with your accepted email address and password.

8 Then enter the PIN you were sent by email, and click Continue. Try and get it right first time, as you only get 3 attempts to enter the correct PIN before failing.

See also

Don't forget those other online payment providers out there, eager for your business. We're big fans of WorldPay (www.worldpay.co.uk), which is part of The Royal Bank of Scotland Group, and is again an excellent choice if you've set up your own business online, and want to accept online payments safely. WorldPay is well respected and has an excellent website which gives you the full lowdown about what's on offer.

Success!

9 You should now be able to continue to the second part of your registration – entering the all-important credit card details. Click on the link to continue your signup.

Did you know?

Nochex is based in Leeds, and has been going since late 1999. Its money payment service has been offer since the beginning of 2001, and it has rapidly grown over the last couple of years to be the primary payment option for many small and medium sized businesses.

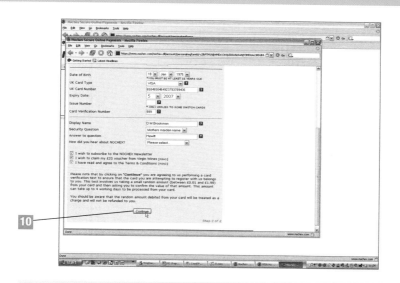

10

Credit card details

10 Now enter all the usual stuff to do with your credit card, including card type, card number, issue number and suchlike. Answer the questions at the bottom of the screen, tick the relevant boxes (especially the I Have Read and Agree to the Terms & Conditions line), then click Continue.

For your information

In the Customer Services area of Nochex, under the Help tab, you can browse a useful knowledge base, or submit a complaint via a ticket system. Presently, there isn't a support telephone number, so we recommend using the tickets to hopefully get a speedier response than simply writing to the Customer Services department in Leeds.

Account types

11 Once you've sorted out all the admin, you're ready to go ahead and use Nochex. We've pointed out a number of useful resources – the Help and Forum tabs really are invaluable before you go diving in anywhere. Also have a look at all the different account types, so that you're well briefed on what the service offers – simply click on Account Types from the home page to get to this at-a-glance rundown of features. Happy trading!

Initially, this end piece was going to be about Nochex Classifieds, a logical extension of the Nochex brand to incorporate online classified ads, which are many people's tips to be a runaway success over the next couple of years. Unfortunately, however, visitors to Nochex classifieds at http://classifieds. nochex.com/ are now greeted with the depressing message: The Nochex Classifieds service has been closed with immediate effect. Our strategic vision is to provide you with a secure, reliable and cost effective payments service, and we believe that Classifieds is not part of this plan. Ouch. So what else does Nochex offer?

Well, click on High Street from the home page to see an impressive list of Nochex merchants and sellers, in categories from antiques to travel. Serious merchants can also apply to become a Nochex partner at www.nochex.com/partners.htm – you'll need to pass some signup criteria to enter this scheme, such as having a level of monthly transactions in excess of £4,000. Pass the test and you can join a plan where you get commission for every successful merchant application that you process through Nochex.

We've mentioned the excellent forums already, and in there, as well as the valuable FAQs and beginner help links, there is a Market Place link, where you can get over the collapse of Nochex Classifieds by finding the odd bargain or two. With over 5,000 members and 50,000 posts, the forums are doing an excellent job of keeping everyone up to speed with all the very latest Nochex developments.

6 eBay PowerTips

WHAT YOU'LL DO

→ **Set up an eBay home page**
→ **Get the most from 'My eBay'**
→ **Join the eBay Community**
→ **Learn about Community etiquette**
→ **Get bargain-hunting tips**
→ **Get PowerSeller tips**

With eBay's massive success, of course, come a few problems for the user. The major question people ask themselves is – how can I become a *better* eBayer? How can I stand out from the crowd, gain precious eBay experience, spot a bargain instantly, and work my way up to the dizzy heights of PowerSeller – the elite band of eBay sellers whose title allows them to boast about being at the very top of the selling tree?

You'll see many, many websites promising to make you a better eBayer, and the bookshelves are groaning with all sorts of takes on the eBay phenomenon. Like so many things in life, there's no 'miracle solution' to becoming a first-class eBay user; there are just certain techniques and plans that you can put into action which make the chances of getting the most out of the site that much higher. That's why this chapter is concerned with 'eBay PowerTips'; tactics that we've learned from our own time on the site and colleagues' experiences, and which will help you force your way up the virtual ladder of top eBayers.

Making use of the tools that are around you is vital, which is why we'll look at the advantages of creating an About Me eBay home page, as well as joining the eBay Community. Mastering My eBay is vital if you want to keep your eBay life in order, and we round things off with 60 buying and selling tips, which don't promise you untold riches, but do point you in some incredibly useful directions for becoming a first-class eBay user. The road to PowerSellership is a tough one, but where would the fun be if everything was easy? And even if you don't make it to PowerSeller status, just putting into practice some of these sellers' expert tactics will improve your eBay life immeasurably, and start getting you noticed by the fellow members of your community.

So, ultimately, you can take out of this chapter as little or as much as you want – maybe dip into it when eBay times are tough, to find a different nugget of information, or alternative tactic which you'd never thought of before. If you're into eBay for the long haul, the wider your knowledge base about the site, the more chances you have of rising to the top.

Want to get noticed as an eBay seller, or just introduce yourself to the community at large? Then you need to create an eBay About Me page.

Accessible as a link from the Community section of eBay, creating an About Me page, whether using plain text or simple HTML computer code, really lets other people know what you're about, and what kind of person you are. It inspires confidence, promotes your listings or eBay Shop, backs up your reputation, and offers people an insight into your world – all key factors in building up that vital trust factor. About Me pages let you list your current auctions, add photographs and pictures, show your recent feedback and generally show off your wares – some people spend hours getting them just right. Your About Me page acts as your very own eBay home page, and its power shouldn't be underestimated; as an overall part of your eBay armoury, it does an extremely effective job. We'll look at the kind of things you can put on your own About Me description over the next few pages.

First moves

1 Click on the Community link from the eBay home page.

2 At the bottom of the Community home page, click on Create An About Me Page.

3 Read all about just what the page can do for you on the next screen. When you've done that, click on Create Your Page.

Timesaver tip

You can engage in a bit of self-promotion on your eBay page. Apart from linking to all your (hopefully) positive feedback, why not give a list of your achievements? We don't mean the 100 m breaststroke when you were six years old, but if a local newspaper has featured your superb eBay selling, for instance, then don't feel afraid to link to it. Anything that makes you stand out from the crowd has got to be good.

Choices, choices

4 You have two different paths to think about next. You can either use the eBay Step-by-Step process to create your page…

5 ...or enter your own HTML code.

6 HTML stands for HyperText Markup Language, and is a system that uses tags to put your text into different kinds of headlines, lists, links and paragraphs. It's essentially a way of telling a browser how it can display images and text, and to this end, it's very useful for creating an eBay home page. You may know no HTML whatsoever, but don't panic – not only are there plenty of places online to learn about it (try www.htmlhelp.com, for example), but eBay has its own guide to useful tags which you can type in when editing your page. Click eBay-Specific Tags.

HTML help

7 You'll be taken to this page, where you can see useful tags for showing your eBay User ID, displaying your feedback, and much more. It takes time to get your head around these tags, but it could be time well spent if it helps the look of your eBay page.

8 Once you've made your choice, it's time for step 2 – the all-important content.

Timesaver tip

Keep eBay's guide to HTML tags bookmarked in your Favourite sites folder, for easy future reference. Its URL is http://pages.ebay.co.uk/help/account/html-tags.html.

All about me

9 Even if you've chosen the step-by-step method, as we have, you can still enter HTML commands – see the line Enter Either Plain Text or HTML.

10 Note that our commands for the moment are simple ones, to do with font style and font size.

11 Paragraph 1 and Paragraph 2 serve to introduce us to eBay users. What are you all about? What's your eBay history? What do you sell? Do you have especially good feedback? Think about answering these sorts of questions in these paragraphs.

12 Be careful as ever about typos. We have a mistake here – I have some of most – which we can easily change once we've spotted it in a preview.

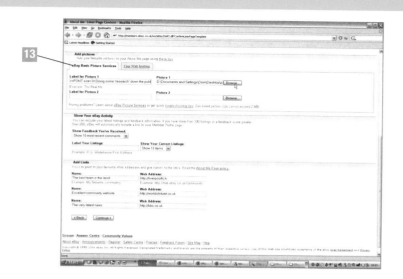

13 Now think about adding pictures to add some personal interest to the page. Think about how many you want, what labels they should have, and whether they add to your page or detract from it. We tried lots of different alternatives.

Timesaver tip

Share the URL that your About Me page is given with friends and colleagues, to make sure that the word spreads quickly.

More info, please

14 Decide whether you want to include your latest listings and feedback information – if you're still a relatively new eBay user, this will be very useful in the future, when you have several listings up and running and hopefully some good feedback.

15 Give people a further insight into your personality by including your favourite web links. Keep 'em clean!

16 On the next page, choose from all the different layout options. Experiment with each one and preview what it achieves – you need to be happy that it's working for you.

Timesaver tip

Keep going back to your About Me page to edit it and keep it fully up to date. It's little use having a home page which goes on about achievements that were years or even months ago. As your eBay life changes over the months, make sure your personal page does as well.

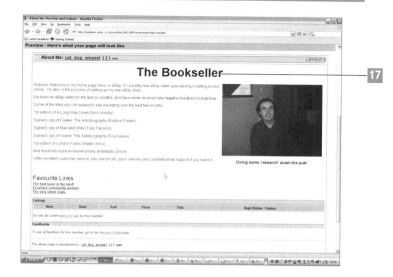

Final steps

17 Here's a very basic page telling the user a little bit about our status and eBay life. Once the listings and feedback comments are added, and we've added a dash of background colour (which you can't see in this black and white book unfortunately!), things will start looking pretty tidy. Check the spelling, formatting, web links, whether you like the photo or not, and everything that goes to make up the page. Remember you need to be updating it frequently, as well.

18 Click Submit to make your page live!

For your information

Tips for the perfect home page

Make your About Me page warm, friendly and accessible, and try to avoid being smug or arrogant. A little bit of personal information about where you come from and what your hobbies are is acceptable, but bear in mind that people are not going to have the time to read your life story. What's more crucial is your relation to eBay – what are you selling? Why should people come to you rather than any other seller? What kind of feedback have you been receiving? If you've clocked up some brilliant feedback, then let people know about it – it will really help people to begin to trust you. Don't shout at people from your page, or litter your writings with childish exclamation marks or spelling and grammar bloopers. Keep it looking as nice as possible, with a little background homework on how HTML works proving immeasurably useful in this respect. Update it frequently if your circumstances change, and remember that the page is one of the most powerful advertising tools you have.

At various times during the course of this book, we've extolled the virtues of hitting the My eBay section of the site to carry out some vital task or admin duty. This area of eBay, accessible from a large tab at the top of your eBay home page, lets you carry out a staggering amount of tasks, including: read your email, manage your Watched Items, manage all your different subscriptions, change your personal account information, use the Selling Manager, sort out your eBay Favourites, look at your Reviews and Guides, report unpaid items, read your recommendations… and much more. Not all of these tasks are massively important or key to your eBay life, but if you take the area as a whole, it's an invaluable way of imposing order on what can be a pretty chaotic eBay existence, especially if you have numerous transactions going on at the same time. Let's dive into its secrets below.

Let's go there

1 You'll probably reach the My eBay area of eBay at lots of different random times when you're using the auction site, but to get there from the home page, just click on the tab at the top of the page.

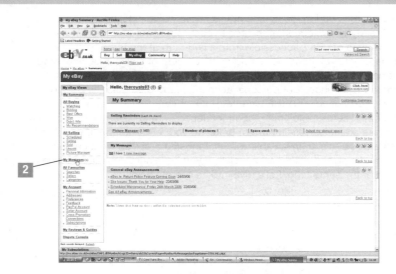

2 You'll see a Summary of the current state of affairs. One of the first things to do could be to check your eBay Inbox – click on My Messages.

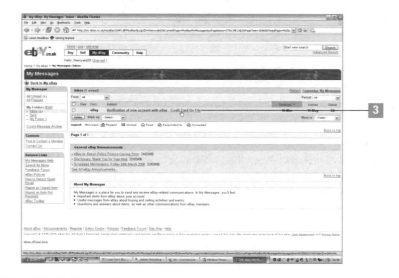

3 We've got a new message. Click on the title of the email to read it.

Timesaver tip

Your Inbox in My eBay is well worth checking pretty regularly, to get all the latest official emails from eBay, confirmations of site action and other general admin messages.

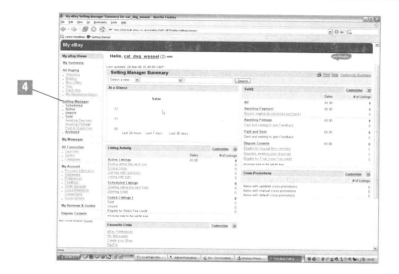

Tweak it

4 If you've signed up for the free Selling Manager tool (see chapter 3), an extra set of Selling Manager links will appear in the left-hand column of My eBay. The Selling Manager gives you your latest selling stats, and shows you the current status of your transactions.

5 If your personal circumstances change, or you just want to access some private information, you need the My Account area of My eBay. Click on its link to explore what it allows you to do.

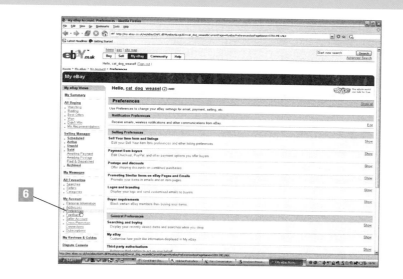

6 Under the Preferences section of My eBay, you can tweak settings such as the payment options you offer buyers, and how information is presented to you when you shop, to your heart's content. Tweak options to get eBay working just how you want it to.

Important

If you change email address, move house or get a different bank account, you really need to remember to go into My eBay and change the corresponding details. Out of date information can ultimately lead to problems which can cause confusion or lead to negative feedback for your capability as a buyer or seller.

Dispute it

7 The Dispute Console plays a vital role in My eBay and your eBay life in general, helping you out when something goes wrong. Its link is on the bottom left-hand side of the screen.

8   It's the place go to if, as a seller, you haven't been paid for the item you've sold, or if, as a buyer, you haven't received an item, or it has arrived, but in a state markedly different from what was described in the listing.

Timesaver tip

Don't go rushing into the Dispute Console without trying other methods of problem management first. The tips page on eBay at http://pages.ebay.co.uk/help/tp/avoid-unpaid-items.html gives you some excellent pointers about how, as a seller, you can try to prevent the dreaded Unpaid Item situation.

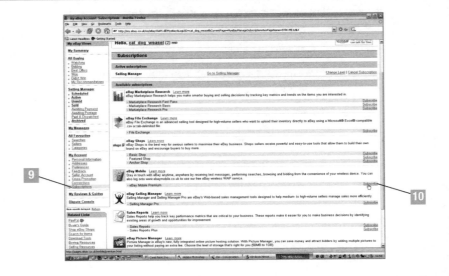

Check subscriptions

9 The more time you spend on eBay, the more difficult it can become to keep check of the status of your subscriptions and what schemes you're a part of. This screen, accessible from the Subscriptions link on the left, lets you keep tabs on the current state of play.

10 Click the Subscribe or Unsubscribe links on the right as appropriate.

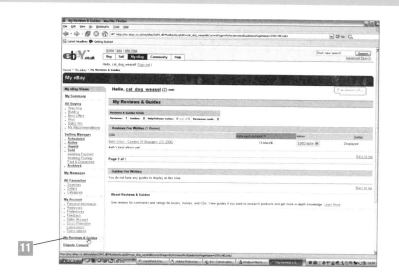

11 Reviews and Guides is an interesting new area of eBay that acts as a useful buying guide – we covered it in chapter 2. Catch up on any reviews you've written, and whether you've achieved a Reviewer rank or not, by clicking the My Reviews & Guides link on the left.

Timesaver tip

Don't think that you can get by on eBay with every detail of your auction life in your head or on random scraps of paper. My eBay is there to help, and making effective use of it will help you become a better eBayer. It lets you keep on top of transactions, deal with problems, update your own personal information and view your bidding history, so it's well worth including in your day-to-day eBay life.

Without getting too quasi-sociological about it all, eBay essentially is one massive community, full of like-minded people interacting with each other through buying and selling. Stories that hit the headlines in the newspapers always seem to be concerned with scams, or disputes, or weird and freaky eBay listings; the more remarkable element of the site is, perhaps, that such a huge group of people co-exist in virtual peace and harmony. One of the factors behind this is the excellent eBay Community area, accessible by one of the large tabs at the top of the eBay home page. The Community lets you discuss eBay-related topics with other users, have a good old gossip and natter about anything and everything, share common interests in different Groups, read the latest announcements and access the impressive Safety Centre. It's an excellent place to pop into from time to time to enhance your eBay experience, so let's see what you can get out of it below.

Hit the Community

1 From eBay's home page, click on the Community tab.

2 A host of options awaits. Potentially one of the most useful is the Feedback Forum.

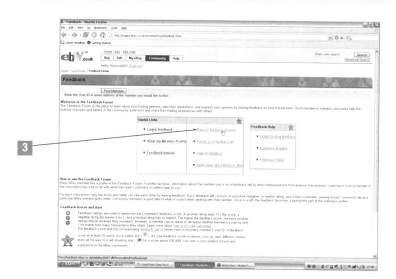

3 As we know, feedback is crucial to the whole ethos of eBay, so understanding it, replying to it and looking at other people's ratings is time well spent. We click on Reply to Feedback Received.

Timesaver tip

We'll deal with some simple chatroom etiquette in the boxout at the end of this task, but if you want the full rundown on acceptable chat behaviour, and what is a definite no-no, view the list at http://pages.ebay.co.uk/help/community/png-board.html.

Time to chat

4　All your feedback appears on screen. To reply to a comment, simply click on the Reply link on the far right.

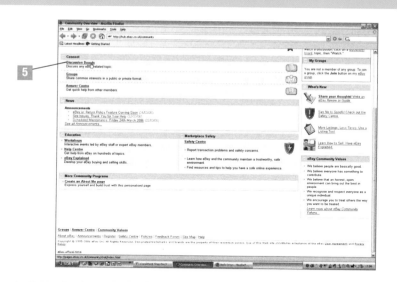

5 Spend time in the feedback forum to really get to grips with how the whole system works. When you're done there, go back to the Community home page and take a look at the Discussion Boards.

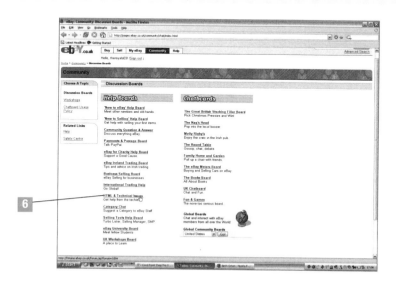

6 A whole host of useful help boards and chatboards awaits. Click on one that interests you.

Timesaver tip

One of the best reasons to reply to feedback is to clarify an important issue, or put your own side politely in opposition to something that has been left about you. If you get a positive comment, replying 'Thanks!' is pretty much a waste of time and rather unnecessary.

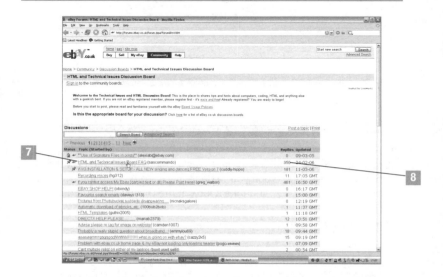

Threads

7 Most of the discussion boards have page after page of topic 'threads' for you to scroll through. Some will be useful, some pointless, so save time by eyeing up the thread headlines to see if they're likely to contain something that could help you out.

8 The number of replies a thread has is a good indication of how useful it could be.

9 The thread we chose is an excellent Frequently Asked Questions topic, where some of people's most common technical problems are dealt with.

10 To add your own two pennies worth to a discussion, click on the Reply link in the top-right corner.

11 Chatboards are a fun alternative to the more serious Help Boards, and a good place to gossip, or pick up vital nuggets of information.

Timesaver tip

eBay Ireland is one of the newest additions to the eBay family. See what Irish trading is all about in the eBay Ireland Trading Board, under 'Help Boards' in the Community.

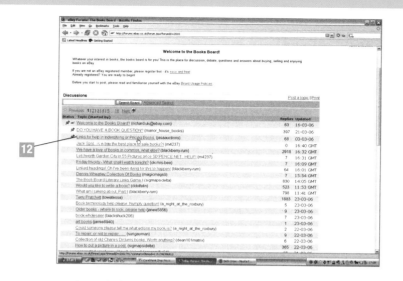

The Books Board

12 We go on The Books Board to prepare ourselves for a life as an eBay bookseller. Tips on here could be invaluable to your future selling life. Don't underestimate the power of research!

13 Use the Chat Boards to your advantage in your on-going quest to become an eBay poweruser. Back on the Community help page, try clicking on Groups if you want to share a common interest with like-minded users.

For your information

Want to know more about eBay Charity auctions, or how you can do your own good deeds for worthwhile causes? Check out the eBay for Charity Help Board to get the lowdown.

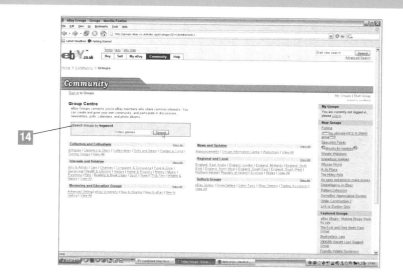

It's good to talk

14 Search groups by keyword to find a group that talks about something you're especially interested in.

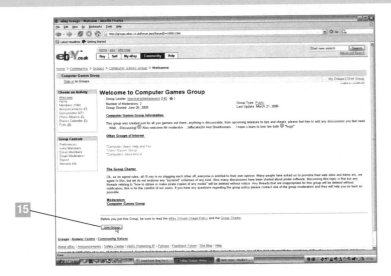

15 We find a video games group very quickly. Click Join Group if you want to join. This isn't just an idle activity – if you're thinking of selling a specific item, talking to other people with similar interests can really help you pick up vital tips. Or, you can just use the Groups to escape from eBay and talk about your favourite hobby!

Timesaver tip

Make good use of the Community as a trouble-shooting device that can help out massively when things go wrong. It's tempting to panic and go in all guns blazing when a negative situation arises on eBay, but if you take time out, look at some of the invaluable resources in eBay Community and stay calm, many of your problems can be solved without having to resort to disputes, threats or contacting eBay.

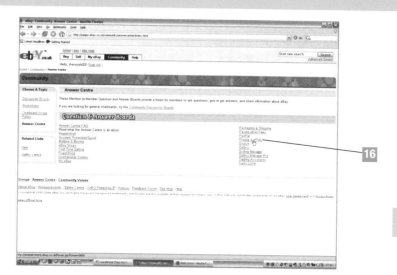

Answers

16 For a quick burst of answers to problems that are bugging you, click on the eBay Answer Centre from the Community home page. The Q&A boards cover a whole host of vital eBay topics, such as PayPal, Packaging and Shipping, TurboLister and dealing with spoof emails.

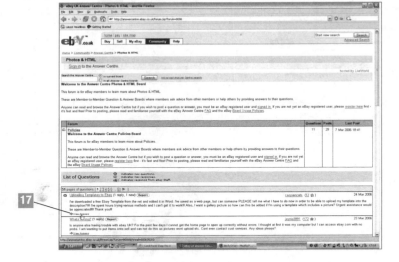

17 Simply read the list of questions on a topic, and click View Answers to see what friendly eBayers are saying in reply.

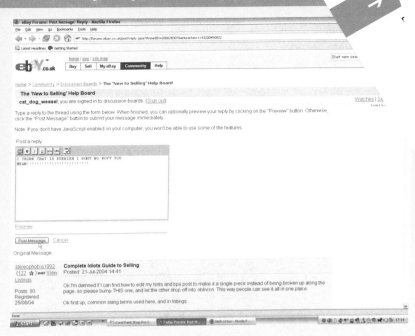

6

If you've been online for a while, chances are that you're used to online chat of one sort or another. Whether you are or not, it's worth reminding you that there is such a thing as 'Netiquette' – how to behave online in chat rooms, messageboards and the like. This screenshot pretty much shows the exact opposite of how you should act when replying to a thread – it's rude, offensive, badly spelt, immature, and ALL IN CAPITALS, which makes it look like you're SHOUTING. Such behaviour is likely to get you in big trouble with eBay moderators, and it's totally unnecessary. Keep your comments to the point, factual, polite and grammatically correct; needless insults will get you nowhere. If you're posing a question, make sure as much as possible that it hasn't been answered elsewhere, either in the forums or the eBay Answer Centre. Asking something that has already been answered countless times will not endear you to your fellow users. Apart from these warnings, enjoy yourself – the community can be an excellent place to either get serious questions answered, or just kick back and shoot the breeze with fellow eBayers.

veryone loves a bargain, of course – the thrill of getting something on the
cheap, or getting something rare, unusual and unavailable anywhere else
online, is one of people's major reasons for returning to eBay time and time
again. How can you spot a bargain quickly, though? How can you be as sure
as possible that the person you're buying from is trustworthy and on the level?
Are there any buying pitfalls to avoid? We'll be answering these questions over
the next few pages, as we give you 30 buying tips to help you get the most
from your eBay shopping. They follow no particular order, so feel free to dip in
and out of them during your regular visits to eBay.

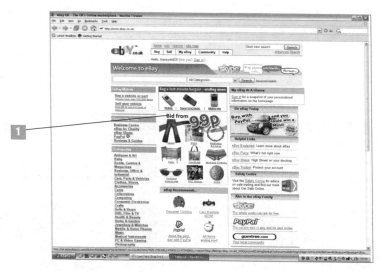

Keep 'em peeled

1 Every day on eBay, you'll see links on the home page to some great-
value auctions that could reap you some fantastic bargains. Look out
for features such as the Bid From 99p boxout.

2 On a similar theme, pay attention to any ending soon auction boxouts on the home page. These allow you to sneak in at the last minute and snap something up, often for low prices, without having to go to the trouble of snapping up a sniping program.

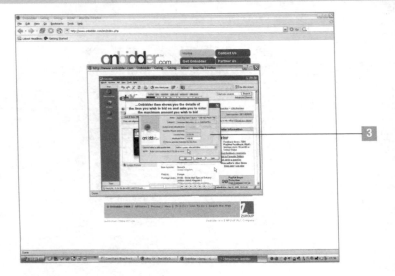

3 If you don't want to rely on this feature, then OnBidder is one of the best sniping programs around, although there are plenty of alternatives on offer. See www.onbidder.com.

For your information

A 12-month subscription to Onbidder will set you back £19.99. The website also has a demo for you to try.

Branching out

4 More and more auctions these days are offering the Buy it Now option as standard, letting you buy instantly without having to risk going through a long drawn-out auction.

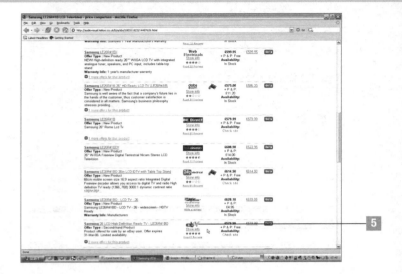

5 Price comparison websites can be a massive help when trying to see whether a price quoted is really such a fantastic deal or not. They're especially useful when, like Kelkoo in this example (www.kelkoo.co.uk), they include eBay prices as part of the comparison.

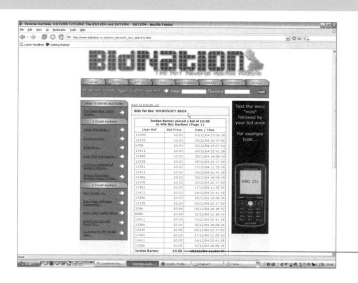

6 Reverse bidding websites really need to be involved in your planning as a buyer, as a useful backup if nothing else. Bidnation (www.bidnation.co.uk) is one of the best – just look at how little people paid to win an Xbox here. Six pence for an Xbox can't be bad!

Important

Remember the major caveat about reverse auctions – you normally have to pay just for the privilege of bidding. So you could end up seriously out of pocket, and not actually win anything.

Trawl your net wide

7 Keep looking at eBay's sister foreign sites as well for great bargains.
You may have the odd language difficulty, but often things are pretty
self-explanatory, and there are plenty of web translation tools around
too (try Babelfish, at http://babelfish.altavista.com/). Here, we're buying
online at eBay France.

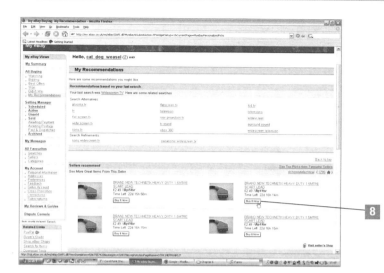

8 Keep checking your recommendations list in My eBay. Every once in a while, something you really want could crop up, often at a bargain price.

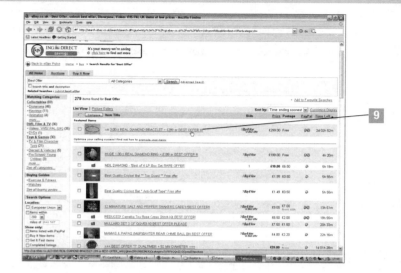

9 Some auctions now offer a Best Offer option, where you as a buyer can suggest a price to buy an item at. The seller can then decide whether to accept or decline your offer, or enter into further negotiations. A bit of bartering can really help you pick up a bargain.

Timesaver tip

To find out more about Best Offers and how they work, check out the help page at http://pages.ebay.co.uk/help/sell/best-offer.html.

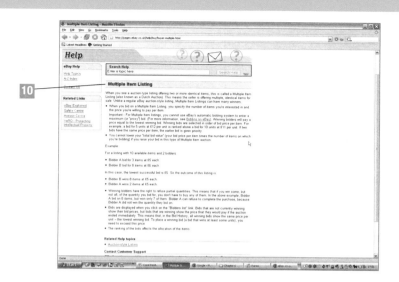

Different kinds of listings

10 Keep your eye open for multiple listings as well, where two or more identical items are offered by the seller. Multiple Item listings can have more than one winner. Get your head around these listings at http://pages.ebay.co.uk/help/buy/buyer-multiple.html.

11 Specialist sites within eBay can offer some brilliant bargains. eBay
Motors (motors.ebay.co.uk) is massively popular, and with some stunning
reductions on what you'd pay for a motor offline, it's not surprising.

12 Never leave eBay Shops (http://stores.ebay.co.uk) out of your online browsing. Many items are on sale in Shops only, so it's vital you factor online stores into your buying considerations. You'll be glad you did.

Timesaver tip

Make sure a motor's eBay listing has plenty of photographs, stacks of information about its history, and the chance to take it for a test drive.

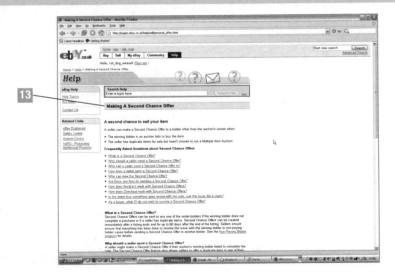

Different offers, community chat and feedback

13 Yet another way of winning an item is through a Second Chance Offer. The chance of doing this will normally arise when an auction has been won by someone, but the winning bidder has failed to buy the item. Again, they offer the chance of picking up a bargain – find out more about why you should know about this form of bid at http://pages.ebay.co.uk/help/sell/personal_offer.html.

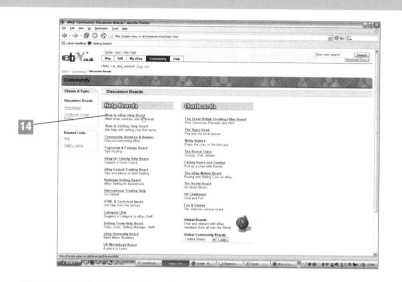

14 A lot of eBayers are so happy with some piece of knowledge they've acquired, that they have to share it with the whole community in one of the Help Boards or Chatboards. Trawling these occasionally can throw up some pure gold nuggets of information that can stop you paying over the odds for something on eBay, and save you a good deal of time and hassle.

15 Ah, the Seller Information box in an auction listing. We must mention it again – this box is key to successful eBay buying. What is their feedback score? Are they a PowerSeller? What's their Positive Feedback ratio? How long have they been a member? What do other eBayers think about him or her? Find out all this crucial information before making a purchase.

Timesaver tip

Read sellers' feedback carefully to check for any skeletons in the closet. The odd mild disagreement or problem can be expected, but if an issue keeps rearing its ugly head from different eBayers about a seller in the feedback list, alarm bells really should start ringing. Do you want to take an unnecessary risk?

Super searches

16 Study the feedback comments carefully. If a seller has lots of different positive comments, from a wide cross-section of users, over a decent period of time, you're on the right track to a happy purchase.

17 An Advanced Search on eBay really helps you zoom in on what you want in your listings results. You can choose to show only Buy it Now items, Get it Fast items, items with the Best Offer option, listings ending within a certain time period.... Cut straight to the chase and remember that a search which yields thousands of results is not necessarily a good search. Time is precious, after all.

18 Another way of saving time (and money, in many cases) is to opt to search for items by a particular seller only. If you've bought successfully from them a few times, then hopefully that will continue when you return to them again and again.

Timesaver tip

We're ramming home how important feedback is to the eBay ethos, so don't forget to leave feedback in every transaction that you carry out on the site. You'll help other users immeasurably and ensure the continuing success and integrity of the site.

Check those spellings

19 Don't forget that you can pick up bargains through looking at all the different spellings for a particular item. It may be annoying that a seller can't be bothered to type something correctly, but if it's a brilliant bargain, and they have good feedback, you may be prepared to forgive them and enter into the auction. FatFingers (www.fatfingers.com) is an excellent 3rd-party utility for digging out mis-spellings. As the site says, other people's typos save you money – many people won't have the presence of mind to be looking at these erroneously-typed listings.

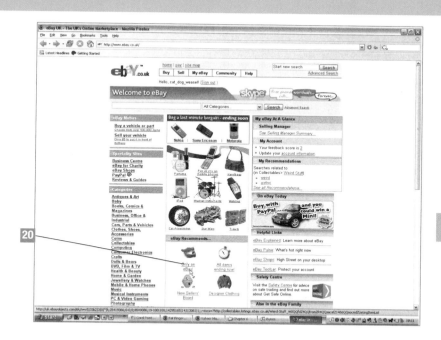

20 Back on the eBay home page, eBay Recommends can help you out in a bargain hunt, too – especially the Only on eBay! link.

21 When any item in real life becomes popular, such as the PSP, Xbox 360 or iPod, expect eBay to be flooded with sales of the must-have. This can work as a double-edged sword, and if you're on the lookout for bargains or exclusives, be careful that an auction isn't asking for crazy money just because the item is unique or not widely on sale in the UK. If you're patient, you can bide your time, look at a number of auctions for the same item in a period of time, and work out who is offering the best deal for that hot item.

Timesaver tip

Try to avoid spelling mistakes in your own listings – it does limit the number of visitors who'll see your listing, and it should only be the work of a few minutes to do a decent spell-check on what you've created.

More tips

22 Does an auction listing tell you everything you need to know? Has important information been left out? If you're left scratching your head, you can email the seller for clarification, but often answers can be unsatisfying. And certainly don't bid for an item where you're not sure exactly what you're getting, or what condition it's in. This book listing has an acceptable photo, and plenty of details in the description which let us know just what we're getting.

23 Photos can be a huge help when it comes to getting a bargain and understanding just what you're getting. Check out the gallery index at http://pages.ebay/co.uk/gallery-index.html.

24 Mobile phone firm O2 has recently been advertising its new i-Mode eBay service, which lets you bid for goods when on the move, with your mobile. This is a great way to stay one step ahead of the game – check out http://i-mode.o2.co.uk/content_whats_new.html for more info.

Timesaver tip

Sellers who not only have excellent feedback, but a well-cared for About Me page, should go to the top of your list. Their About Me page may also point you in the direction of some superb bargains that they're currently offering.

6

25

Alternatives

25 Have a decent collection of watched auctions in My eBay, although not too many, as you won't be able to keep track of them all. Follow an auction that you're interested in closely, look at the bidding patterns, and then strike when you think the iron is hot.

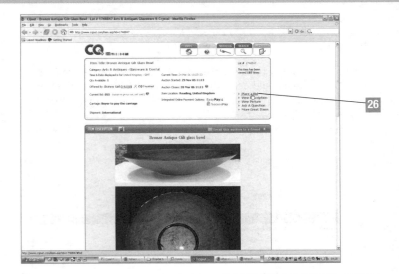

26 Other auction sites really do need to be on your radar, as the best ones can offer you the same thing that you're interested in on eBay, but cheaper. CQOut (www.cqout.com) is one excellent alternative.

27 As is QXL (www.qxl.com). eBay may be your number one, but any bargain hunter worth their salt looks in more than one place.

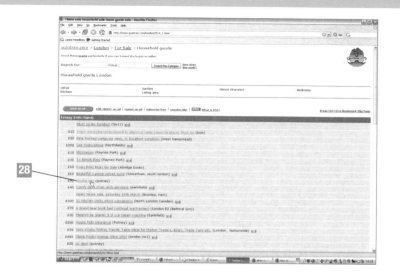

Final tips

28 eBay's partner Gumtree has superb bargains wrapped up in its classified ads. Take a looks as you may well be surprised at the range of items on offer at rock-bottom prices.

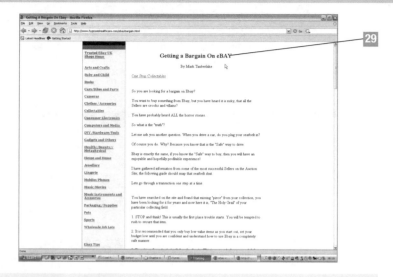

29 Look for well-written bargain tips on the internet as well, such as this effort at www.hypnosishealthcare.com/ebaybargain.html.

30 Don't forget to check out the Featured Items when you're buying on eBay. You'll in all likelihood get an eclectic mix of items, but there could just be the bargain of the century lying in amongst the things you have no intention of buying.

eBay describes PowerSellers as 'exemplary sellers who are held to the highest standards of professionalism, having achieved and maintained a 98% positive feedback rating and an excellent sales performance record'. Membership to this exclusive club is free, but the qualifications are high, as http://pages.ebay. co.uk/services/buyandsell/powerseller/criteria.html points out. Requirements include averaging a minimum of £750 in sales per month, for three consecutive months, achieving a feedback rating of 100, of which at least 98 per cent of which is positive... and then maintaining your ratings every month, when your membership comes up for renewal. You may think that the high volume of sales that are required to become a PowerSeller puts the badge out of your reach; even if this is the case, however, you can still adopt some first-class PowerSeller-esque tactics to make your selling life as pleasurable and profitable as possible. We'll be looking at these kinds of tactics over the next few pages; there's no guarantee that they'll turn you into a PowerSeller, but they will help your sales life immeasurably, and if you combine the tactics with a high volume of sales, that exclusive club may start appearing eminently achievable after all.

6

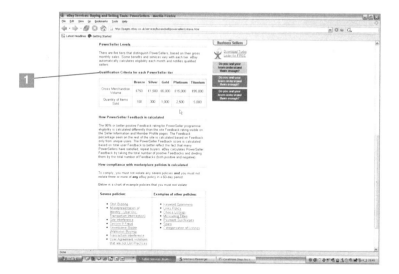

How do I do it?

1 If you're going to try and become a PowerSeller, you need to know just what the requirements are. It's a steep task, and there're five different tiers of PowerSeller as well. Take a look at what's expected of you at http://pages.ebay.co.uk/services/buyandsell/powerseller/criteria.html.

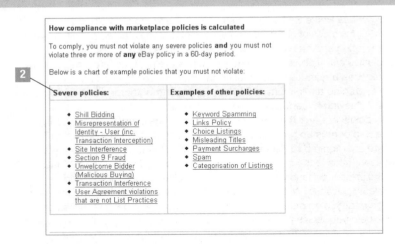

How compliance with marketplace policies is calculated

To comply, you must not violate any severe policies **and** you must not violate three or more of **any** eBay policy in a 60-day period.

Below is a chart of example policies that you must not violate:

Severe policies:	Examples of other policies:
• Shill Bidding • Misrepresentation of Identity - User (inc. Transaction Interception) • Site Interference • Section 9 Fraud • Unwelcome Bidder (Malicious Buying) • Transaction Interference • User Agreement violations that are not List Practices	• Keyword Spamming • Links Policy • Choice Listings • Misleading Titles • Payment Surcharges • Spam • Categorisation of Listings

2 Understand what violations of site policy could cost you PowerSeller status in the list of example policies that you must not violate, on the same page.

3 One of the key aspects of becoming a PowerSeller is, of course, getting that all-important feedback rating into the stratosphere. A couple of negative comments are allowed, as in this PowerSeller's case, but you need to strive for the utmost standards in quality listings, postage and delivery and customer care to even get close to becoming part of the eBay elite.

Selling tips

4 In My eBay, My Account, you can see details of any outstanding
invoices that need paying. Prompt payment of invoices will help in the
quest for PowerSeller status.

5 As a seller, you need to have a firm grasp of what people actually want.
All the way through a typical year, for example, new gadgets and
gizmos will suddenly become hot – understanding this and getting in on
the action fast can make all the difference to your seller status. Here,
we can see sellers who've been quick off the mark selling Nintendo's
hot new gadget, the Nintendo DS Lite. The item's limited availability will
make auctions for it all the more likely to have some frenetic action.

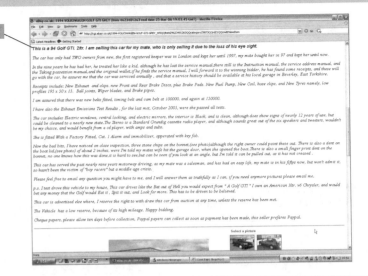

6 If there are things you need to say in your listing, say them. Don't cut corners or be vague and inaccurate. There's no need to waffle, but detailed descriptions such as this one, listing a 94 Golf GTi, are ideal and really impress buyers.

Timesaver tip

Think about maybe setting up a Direct Debit account to help you get those all-important fees and invoices paid promptly.

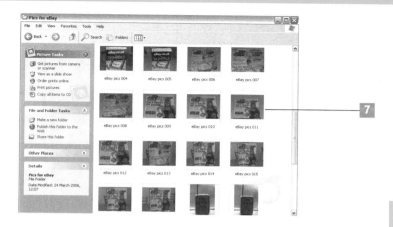

Picture perfect

7 Don't skimp on your item photos. Have a folder on your PC full of your pictures, transferred from a digital camera. Have a wide selection of photos of the same item – that way you can discard the dodgy photos that didn't come out right, and keep the best-looking photos.

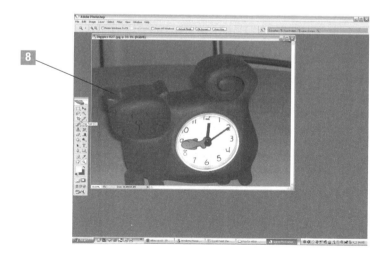

8 When you find a picture that you're happy with, use an editing program such as Adobe Photoshop to crop away any extraneous detail, and get your photo in as presentable a shape as possible.

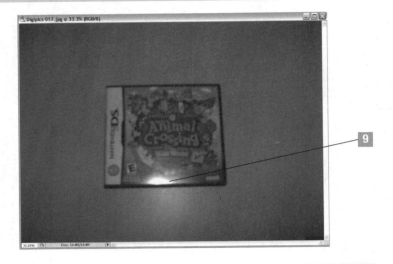

9 If you take a poor photo, go back and take it again! Don't include a sub-standard image, as it can really take away from the quality of your listing. It certainly won't impress buyers. Our picture of a Nintendo DS game here is blurry and has a horrible lighting glare on it – not impressive at all. Bin it!

Timesaver tip

There are plenty of beginner-level photography magazines and books on the shelves for you to browse through and get camera tips. You don't need to be some sort of David Bailey-esque character to take photos of your items, but a clear understanding of some of the basic principles of photography can be a real help.

Seller options

10 Offer the most flexible posting options that you can to attract buyers from far and wide.

11 Offer some incredible bargains to tempt buyers. This P2 game could be about to go for 99p, plus postage and packing.

12 As a seller you can select some Buyer Requirements when you list an
item. Be careful, as you obviously don't want to limit the number of
your auction's potential visitors more than necessary. If you want to try
and limit the chance of getting a dodgy buyer, you can block buyers
with a negative feedback score. Remember, however, that some people
can get negative feedback unfairly, so by clicking this option, you could
be just blocking people who've been unlucky or victims of misfortune.

Timesaver tip

Think carefully about the postage costs you set in your listing. You
don't want to end up out of pocket, but at the same time you don't
want to put off potential buyers with a shoddy, slow delivery service.

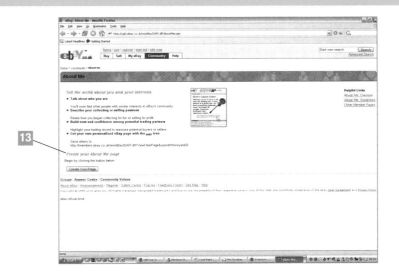

More tips

13 Time spent creating an About Me page can really help your reputation
as a seller. Think carefully about spending a couple of hours getting one
up and running – it could really make the difference to your sales.

14 A proper eBay shop can look incredible. Just look at this store, Candlecity at http://stores.ebay.co.uk/Candle-City-Limited. Very impressive.

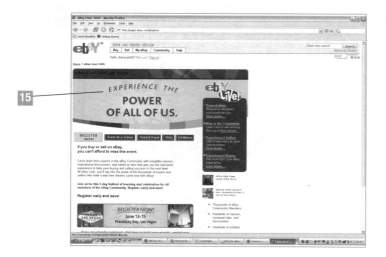

15 eBay Live is an event held every year, normally in glitzy places such as Las Vegas. It's a chance for eBayers to gather together, learn from the experts, and generally take their buying and selling life to a whole new level. See if it's something you want to make time for by going to http:pages.ebay.com/ebaylive.

Timesaver tip

Setting up an About Me page also adds a 'me' icon next to your name in eBay, so people can see what you're really about.

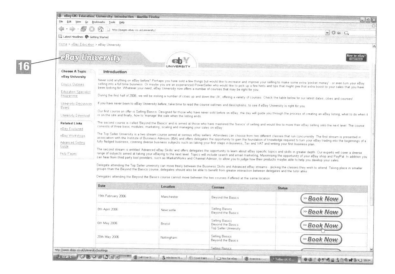

Back to school

On a similar theme, the travelling eBay University across the UK can help people of all levels of eBay skill become even more advanced. Dates are held pretty regularly throughout the year in the UK – check out http://pages.ebay.co.uk/university.

17 If buyers ask you questions about your listing on eBay, take the time out to answer them promptly and courteously, dealing with precisely the issue in hand.

18 As your eBay experience grows, there's a chance you may want to change your username – maybe from a personal one to a business-related title, if you're starting to sell in large volume. Think carefully about what your User ID says about you. You can change it from the My eBay area of eBay.

Timesaver tip

If you're an eBay shopowner, you really should think about having a User ID that accurately and meaningfully reflects the nature of your business.

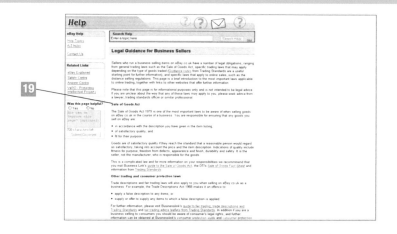

Know where you stand

19 If you're serious about running an eBay business, check out the legal ramifications at http://pages.ebay.co.uk/help/policies/business.html.

20 Confident that your eBay business is getting off the ground? Then why not build a separate website? You can then promote your auctions on your own site, thus, hopefully, significantly increasing the number of people visiting your listings.

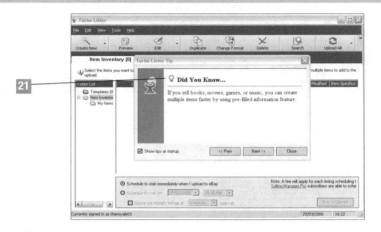

21 Make full use of eBay's Turbo Lister – it really is one of the best free utilities out there when it comes to quickly getting professional-looking listings set up. Simply create listings offline, then use the power of the program to get them all uploaded to eBay instantly, saving you countless hours of hassle.

Timesaver tip

Keep your eyes peeled in your eBay Inbox for any add-ons or refinements to Turbo Lister, which get carried out by the utility's programmers from time to time.

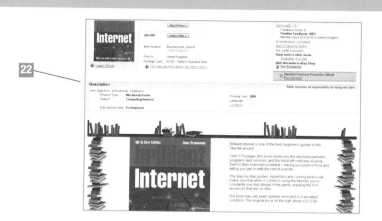

Power tips

22 When listing, make full use of the Preview option before submitting your auction listing. That way, you can make sure you're 100 per cent happy about what you've done, and can iron out any niggling little spelling or grammar mistakes.

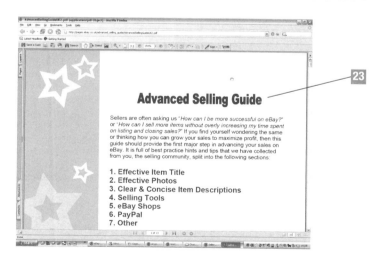

23 eBay has its own PDF Advanced Selling guide which is well worth a look when it comes to trying to be a better seller. Take a look at http://pages.ebay.co.uk/advanced_selling_guide/AdvancedSelling GuideUk2.pdf.

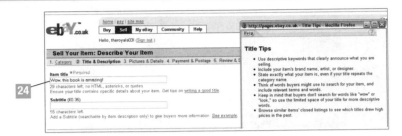

24 The title of your eBay listing is vital. Look at what we've written here – apart from telling us it's a book, the title tells us absolutely nothing of any relevance or importance. Check out eBay's title tips to avoid howlers such as this title – stick to the basics of exactly what your item is, and don't try to be clever.

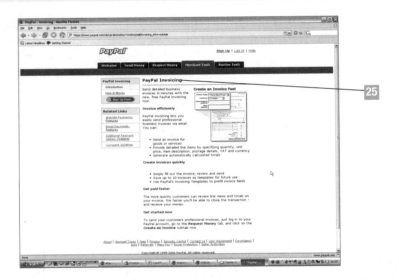

Software options

25 Remember, effective use of PayPal can be a real boon in your selling life. Not only is it a great way of offering buyers hassle-free payment avenues, but it also has its own range of tools which can really help your sales, such as this free PayPal invoicing tool. Its under Merchant Tools at www.paypal.com/uk.

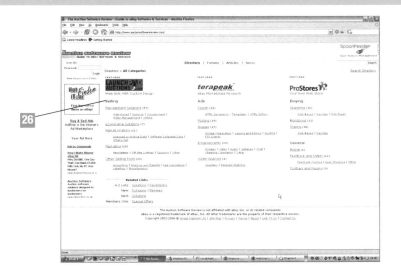

26 Carry out the occasional Google search to see what useful auction software is on the market – there are a lot of free trials and free utilities that mean you don't have to spend a penny to get something really useful. Here we're at The Auction Software Review (www.auctionsoftwarereview.com).

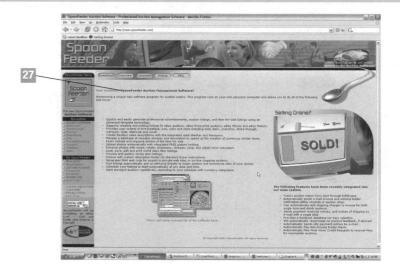

27 SpoonFeeder Auction Software (www.spoonfeeder.com) is another utility you should be aware of. Check out the program's benefits on the site's home page.

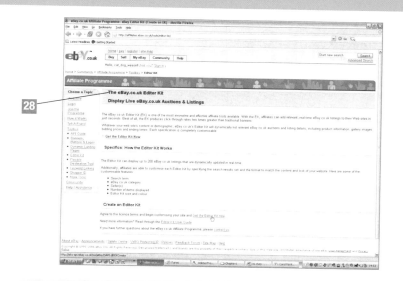

Final tips

28 The eBay Editor Kit (http://affiliates.ebay.co.uk/tools/editor-kit) lets you add relevant eBay auctions to your own web site quickly and easily. See if it's something you could be interested in at the above URL.

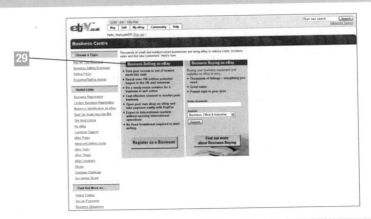

29 The eBay Business Centre (http://pages.ebay.co.uk/businesscentre/index.html) can really help your fledgling business stimulate sales and make it to the top.

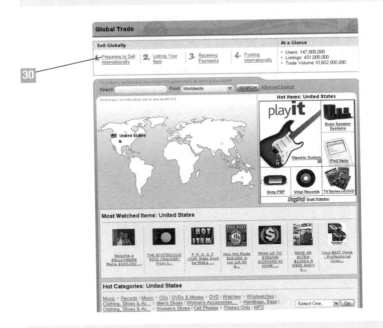

30 And don't forget the global market when it comes to being a seller. If you're someone who can sell further afield than just your own back yard, you'll be increasing your sales potential massively.

For your information

Examples of Power Sellers

Obviously you can check out PowerSellers by clicking on the PowerSeller logo next to their username in an auction listing. It's important to remember that PowerSellers are under the constant monthly pressure of their status being reviewed; so if someone has been a PowerSeller for a long time, that's a huge indication that they're a terrific seller. PowerSellers tend to be quite a secretive bunch, understandably protective of their elevated status which they've worked so hard to achieve. As a buyer, when you're dealing with someone who's a PowerSeller, you can be assured that the transaction will go as smoothly as possible . Check out eBay's monthly featured PowerSeller at http://pages.ebay.co.uk/services/buyandsell/powerseller/superps.html.

6

For your information

More money-making tips

Treat fellow users courteously, offer them good value on useful items, make your listings thorough yet concise and relevant, deal with queries immediately, and offer brilliant postage and shipping. Don't forget to think about selling your items on alternative auction sites to eBay, if you have the time and inclination. If you're really serious about making some significant piles of cash, you're going to need to put some time and effort into getting a first-rate shop online, and maybe an external website as well.

Jargon buster

About Me – Create an About Me page from the Community area in eBay, to tell fellow users all about your selling accomplishments. It's one of the best promotional options around.

Adobe Photoshop – An image-editing program that can help you get your eBay pictures in tip-top condition. Its younger, cheaper cousin is called Photoshop Elements.

Amazon Marketplace – Spin-off from the main Amazon site, which lets you sell your goods to Amazon customers and make money.

Buy it Now – Buy items quickly on eBay at a fixed price, without having to go through the uncertainty of an auction.

Categories – Items on eBay are grouped into their relevant categories, such as Baby, Computing and Crafts. Putting your item in the right category when you list on eBay is vitally important.

Chat room – A special place on the web where you can go and talk to like-minded users about virtually any topic you can think of. eBay has plenty of chat rooms where you can gossip or talk about serious auction-related issues.

Checkout – eBay Checkout is a set of tools provided by the auction giant to smooth out the process of completing a transaction with another user.

Dispute Console – The Dispute Console, accessible from My eBay, lets you file a complaint about an unpaid item or an item that you haven't received yet, or one which came in a markedly different state than the auction said it would.

eBay Pulse – eBay Pulse is the regularly updated guide to the zeitgeist of eBay, telling you what the most popular current searches are, and what's hot right now.

eCommerce – A general term for conducting business or transactions online.

Feedback – The crucial system underlining how eBay works. When you take part in a transaction with another eBay user, you should leave feedback to say how the deal worked. Users can then build up feedback ratings which let others know what kind of eBayer they would be dealing with.

Fees – eBay has a number of fees that you need to take into account when using the site. Take a look at http://pages.ebay.co.uk/help/sell/fees.html.

Gallery – Get your item featured in the eBay Gallery (at a cost) to give buyers a quick visual clue about what you're selling.

Gumtree – A successful online classified ads site, recently bought by eBay.

HTML – Stands for HyperText Markup Language. The universal language of the web, in which all pages are written. Use HTML in your eBay home page to impress visitors.

Links – Links take you around the web. By clicking on a link, you're taken to a different website or page.

Listing – When you put an item up for sale on eBay, you're creating a listing for all of eBay's users to see.

My eBay – A crucial area of eBay, where you can carry out vital admin, and see your current buying and selling state of affairs.

Netiquette – A way of behaving online, especially in relation to chat rooms, newsgroups and messageboards. Being polite, courteous, relevant and to the point is good Netiquette.

Password – A combination of words and numbers, personal to yourself, that lets you access a particular web service such as eBay.

PayPal – eBay's sister online payment service, used by millions of users across the world. It's one of the best ways to carry out online transactions, such as paying an eBay seller.

Phishing – An internet scam where a fake email is sent to you, often purporting to be from a large company such as eBay, trying to con you into revealing personal account details and passwords. Once phishers get this information, they use it to act fraudulently and essentially steal money from you.

Reverse Bidding – A great new way of bidding online. Reverse bidding falls into a couple of different formats, the most common being where you pay to make a bid, and then see if your offer becomes the lowest unique bid in the process. If it does, you can get some incredible bargains, such as a car for just a few pence, or a luxury holiday for a pound.

Reviews and Guides – eBay's new area encouraging users to rate products such as CDs, DVDs and games, and also provide other users with a guide to buying in a particular category.

Search engine – A powerful resource such as Google (www.google.co.uk) which helps you find websites and internet content quickly.

Second Chance – A seller can make a Second Chance offer to someone other than the auction's actual winner, if the winner fails to buy the item.

Selling Manager – A very useful free tool from eBay, which lets you give your selling life a bit more order and efficiency.

Skype – Another service recently bought by eBay, which lets you chat to millions of users across the globe at rock-bottom prices.

Sniping – A controversial facet of eBay life, where people use 'sniping' programs to bid in the very-last seconds of an auction, theoretically increasing their chances of winning the item.

Subscriptions – When you sign up to features such as eBay Shops, you're taking out a subscription that has a monthly charge. Check out the status of all your subscriptions from My eBay.

Toolbar – A bar of buttons that normally sits at the top of your browser window, helping you use a certain program. Popular toolbars online include the Yahoo! toolbar, Google's toolbar and eBay's toolbar (http://pages.ebay.com/ebay_toolbar).

Turbo Lister – A great downloadable tool from eBay, which lets you create your listings offline before uploading them all instantly to the site.

Upload – The technical term for copying all your text and images to the web.

Username – When you register with eBay, choose a good username which reflects your name or what you're going to be selling. All other eBayers will see your username when they come to deal with you.

Watching – If you're keen to follow the progress of an item on eBay, 'watch' it. This allows you to get updates on the auction's progress via email and via the My eBay section of the site.

Troubleshooting guide